Engineering Production-Grade Shiny Apps

Chapman & Hall/CRC
The R Series

Series Editors

John M. Chambers, Department of Statistics, Stanford University, California, USA
Torsten Hothorn, Division of Biostatistics, University of Zurich, Switzerland
Duncan Temple Lang, Department of Statistics, University of California, Davis, USA
Hadley Wickham, RStudio, Boston, Massachusetts, USA

Recently Published Titles

Learn R: As a Language
Pedro J. Aphalo

Using R for Modelling and Quantitative Methods in Fisheries
Malcolm Haddon

R For Political Data Science: A Practical Guide
Francisco Urdinez and Andres Cruz

R Markdown Cookbook
Yihui Xie, Christophe Dervieux, and Emily Riederer

Learning Microeconometrics with R
Christopher P. Adams

R for Conservation and Development Projects: A Primer for Practitioners
Nathan Whitmore

Using R for Bayesian Spatial and Spatio-Temporal Health Modeling
Andrew B. Lawson

Engineering Production-Grade Shiny Apps
Colin Fay, Sébastien Rochette, Vincent Guyader, and Cervan Girard

Javascript for R
John Coene

Advanced R Solutions
Malte Grosser, Henning Bumann, and Hadley Wickham

For more information about this series, please visit: https://www.crcpress.com/Chapman--HallCRC-The-R-Series/book-series/CRCTHERSER

Engineering Production-Grade Shiny Apps

Colin Fay
Sébastien Rochette
Vincent Guyader
Cervan Girard

CRC Press
Taylor & Francis Group
Boca Raton London New York

CRC Press is an imprint of the
Taylor & Francis Group, an **informa** business

A CHAPMAN & HALL BOOK

First edition published 2022
by CRC Press
6000 Broken Sound Parkway NW, Suite 300, Boca Raton, FL 33487-2742

and by CRC Press
2 Park Square, Milton Park, Abingdon, Oxon, OX14 4RN

Library of Congress Cataloging-in-Publication Data

Names: Fay, Colin, author. | Guyader, Vincent, author. | Rochette, Sebastien, author. | Cervan, Girard, author.
Title: Engineering production-grade shiny apps / Colin Fay, Sébastien Rochette, Vincent Guyader, Girard Cervan.
Description: First edition. | Boca Raton : CRC Press, 2021. | Series: R series | Includes bibliographical references and index.
Identifiers: LCCN 2020056843 (print) | LCCN 2020056844 (ebook) | ISBN 9780367444679 (hardback) | ISBN 9780367466022 (paperback) | ISBN 9781003029878 (ebook)
Subjects: LCSH: Web applications--Development. | Software engineering--Management. | R (Computer program language)
Classification: LCC QA76.76.A65 F395 2021 (print) | LCC QA76.76.A65 (ebook) | DDC 006.7/8--dc23
LC record available at https://lccn.loc.gov/2020056843
LC ebook record available at https://lccn.loc.gov/2020056844

ISBN: 978-0-367-44467-9 (hbk)
ISBN: 978-0-367-46602-2 (pbk)
ISBN: 978-1-003-02987-8 (ebk)

DOI: 10.1201/9781003029878

Typeset in Latin font
by KnowledgeWorks Global Ltd.

To our families,

who still think we fix printers and broken WiFi.

To Diane,

thank you for always believing in us.

Contents

V Step 4: Strengthen   161

11 Build Yourself a Safety Net   163

12 Version Control   195

VI Step 5: Deploy   205

List of Figures

Introduction

Motivation

This book will not **get you started with {shiny}**, nor **talk how to work with {shiny} once it is sent to production**. What we will be discussing in this book is **the process of building an application that will later be sent to production.**

Why this topic? Lots of blog posts and books talk about getting started with {shiny} (Chang et al., 2021a) or about what to do once your application is ready to be sent to production. Very few (if any) talk about this area in-between: how to confidently work with {shiny} once you know the basics, and before you send it to production.

This is what this book is going to talk about: building robust {shiny} applications that are ready to be sent to production. We'll focus on the process, the workflow, and the necessary tools for building production-grade {shiny} applications.

Audience for this book

If you are starting to read this book, we assume you have a working knowledge of how to build a small application using {shiny}, and want to know how to go one step further. In other words, you already have some knowledge about how {shiny} works, are able to build a small working application, and want to know how to build a serious, production-grade application that can be sent to production.

The content of this book oscillates between conceptual discussions (e.g., what is complexity?), exploration of project-management questions (e.g., how do we organize work when building production {shiny} applications?), and technical topics (e.g. what are {shiny} modules, or how do I optimize {shiny}?), with a large portion of the book being dedicated to technical questions.

This book will then be of interest to two groups:

- Team managers who want to help to organize work, and {shiny} developers who want to learn about project management. This group will find relevant content in the first 1/3 of this book (roughly until Chapter 5).

- Developers who want to cover medium to advanced {shiny} topics that will be relevant to production. This group will be more interested in the rest of the book, which roughly spans from Chapter 6 to the end. Note that the technical elements covered in this book range from intermediate to advanced topics, and we hope that you will find relevant topics for your current goals, whatever your level is.

In practice, we've come to realize that a lot of {shiny} projects are built by small teams with a large number of these "teams" being composed of only one developer. In this context, we have realized that project management and development are often assigned to the very same person,[1] making these two skills deeply integrated when carrying production projects.

That's why this book tries to reconcile both these worlds, as building production-grade software needs a reliable and operational code-base which is made possible by having solid technical skills, but also reliable and operational team work and project management.

What is "production"?

All throughout this book, we will be using the term "production" to refer to the way we build and deploy our application. But what is "production"?

There have been a lot of definitions of what production is, and even today if you ask around you will get a wide range of different answers. We like to think that a piece of software is in production when it combines the three following properties, not only for the users, but also the engineers working on it:

- It is **used**, even if only by one person.

- It is **relied upon** by its user(s).

- It has **real life impact** if something goes wrong.

---

[1] Even if there are several developers on the team, the project manager is also involved in the coding process.

These three properties impact two specific groups: users and the developers.

Indeed, the users rely on the app to work so that they can do their job, and expect it to deliver meaningful results that they can count on. From the engineering point of view, a production-grade software can be relied upon in the sense that developers count on it to run as expected, and they need to rely on the software to be resilient to change, *i.e* to be modular, documented, and strongly tested so that changes can be integrated with confidence.

A production software also has real-life impact if something goes wrong: users will make wrong decisions, they might be unable to do their day-to-day work, and there are all the things that can happen when the software you use on a daily basis fails to run. From the engineering point of view, a production-grade software has real impact when something goes wrong: someone has to fix the bug, the company selling the software might lose money, data can be lost, and so on.

Given these two properties, you can understand why being in production doesn't necessarily mean being served to gazillions of users,[2] and serving trillions of gigabytes of data: even software that is used by one person who relies on this application to do their job is a production software.

This is what this book is about: building {shiny} applications that can be used, on which you and your users can rely, and including all the tools that will help you prevent things from going wrong, and when they eventually do, making sure you are equipped to quickly fix the bugs.

Book structure

- Part 1, "Building Successful {shiny} Apps" gives a **general context for what we mean by "production-grade" and "successful" {shiny} applications, and what challenges arise when you are dealing with a large-scale application designed for production.** In this part, we will define what we mean by "Successful", stress the importance of project management, develop how to structure your project for production, and introduce the {golem} (Fay et al., 2021) package. We will finally, briefly introduce to our development workflow: a workflow that will be explored in Parts 2 to 6.

- Part 2 through 6 **explore the workflow for building successful applications.**
  Part 2: Design (Step 1) underlines the centrality of the user experience

---

[2]"Production" being equal to tons of users is a definition we regularly hear.

when engineering an application, and emphasizes the importance of designing before coding.

Part 3: Prototype (Step 2) stresses the importance of prototyping, explores the setting of a {golem}-based application, and presents {shinipsum}, {fakir}, and the "Rmd First" development methodology.

Part 4: Build (Step 3) explores the building step of the application, *i.e* the core engineering of the application once the prototyping phase is finished.

Part 5: Strengthen (Step 4) explores in-depth testing, continuous integration, and version control.

Part 6: Deploy (Step 5) develops the various possibilities for preparing your application to be deployed.

- Part 7, "Optimizing" **tackles the question of optimization**, first by introducing the general concepts about optimization, then by presenting some common caveats of {shiny} apps, and then showing how to optimize R code, and how to use JavaScript to lighten R work. This part ends with a gentle introduction to CSS (Cascading Style Sheets).

About the authors

Colin Fay

Colin has written the vast majority of this book. He's responsible for its general structure, and for the original design of the workflow described. Most of the time (if not every time) "we" actually refers to him. He is the lead developer of the {golem} framework, and creator of many tools described in this book.

Colin works at ThinkR, a French agency focused on everything R related. During the day, he helps companies to take full advantage of the power of R, by building tools (packages, web apps, etc.) and setting up infrastructure. His main areas of expertise are data and software engineering, infrastructure, web applications (front-end and back-end), and R in production.

During the night, Colin is also a hyperactive open-source developer and an open-data advocate. You can find a lot of his work on his GitHub account (https://github.com/ColinFay) and on ThinkR's account (https://github.com/thinkr-open).

He is also active in the R and Data community, and is an international speaker.

Sébastien Rochette

Sébastien has been instrumental in the review of most of this book's chapters. He has also written the section about prototyping in RMarkdown, a concept he initiated.

Sébastien is a data scientist at ThinkR, where he teaches anything R related from beginner to expert level, guides R developers towards implementation of best practices, and creates tailor-made R solutions for the needs of his customers.

Vincent Guyader

Vincent is the founder of ThinkR. He created the first proof-of-concept framework for {shiny} applications inside packages; an idea which has led to the creation of {golem}. If you feel like a GitHub archaeologist, this very first version is still available with a little bit of exploration!

With more than ten years of experience with R, and a scientific and technical background, Vincent is an R enthusiast. He still has his hands in the code, whether to develop applications, analyze data, or build packages. When he's not coding, he plays with Docker and manages servers. He strongly believes that meeting highly technical challenges is not incompatible with pedagogy: he passionately trains very diverse learner profiles in R.

Cervan Girard

Cervan has worked on some of the example applications that are used inside this book, namely {shinipsumdemo}, {databasedemo}, {grayscale}, {bs4dashdemo}, and {shinyfuture}.

Cervan is a Data Scientist at ThinkR. He is enthusiastic and motivated when it comes to rolling up his sleeves for new challenges, even if it means venturing dangerously into the depths of R, learning new languages, and experimenting outside your comfort zone.

Whatever the challenge, he remains reliable, constructive, and efficient when it comes to using his skills to train or develop. He also enjoys training learners of all levels in the R language.

Disclaimer

Open source is moving (very) fast, and some of the tools described in this book are still under active development. But good news! **A large portion of this book is about the methodology, and not purely the technology**, so even if some of the packages and code sections used in this book can (and will) become obsolete, a significant part of what is described in this book will still be relevant.

When reading this book, remember that they are a "snapshot" of various technologies and packages, which might someday become out of date, have breaking changes, or even disappear. The last revisions of the pages you're reading now have been done on July 16, 2021. We will **try to update the online version whenever changes are made to the packages used in this book**, so feel free to browse the online version[3] for up-to-date information.

Software information and conventions

This book was built with {knitr} (Xie, 2021b) and {bookdown} (Xie, 2021a).

Package names are in curly brackets in code format (e.g., {rmarkdown}), and inline code and file names are formatted in a typewriter font (e.g., knitr::knit('doc.Rmd')). Function names are formatted in a typewriter font and followed by parentheses (e.g., render_book()). Larger code blocks are formatted in a typewriter font and have a gray backgroud, e.g.:

```
install.package("golem")
```

When we describe a package that can be installed from GitHub, we use the install_github() function from the {remotes} (Hester et al., 2021) package. In other words, the following code:

```
remotes::install_github("ColinFay/dockerstats")
```

---

[3]Available at https://engineering-shiny.org.

means that the package will be installed from GitHub, and that you can use the {remotes} package, which has to be installed on your machine if not already there.

Want to help?

Any feedback on the book is very welcome. Feel free to open an issue[4], or to make a Pull Request if you spot a typo.

Other resources

One single book could not cover everything there is to know about {shiny}. Here are some resources you can use in addition to this book. You can also find more resources in the bibliography.

Getting started with {shiny}

- Mastering Shiny[5]
- Learn Shiny with RStudio[6]
- Getting Started with Shiny[7]
- A gRadual intRoduction to Shiny[8]
- Shiny From Start To Finish[9]

Putting {shiny} into production

- Shiny in production: Principles, practices, and tools[10]
- Shiny in Production[11]

---

[4]https://github.com/ThinkR-open/building-shiny-apps-workflow/issues
[5]https://github.com/hadley/shiny-book
[6]https://shiny.rstudio.com/tutorial/
[7]https://ourcodingclub.github.io/2017/03/07/shiny.html
[8]https://laderast.github.io/gradual_shiny/
[9]https://github.com/rstudio-conf-2020/shiny-start-finish
[10]https://resources.rstudio.com/rstudio-conf-2019/shiny-in-production-principles-practices-and-tools-joe-cheng
[11]https://kellobri.github.io/shiny-prod-book/

Feel free to suggest a resource[12]!

Acknowledgments

Special thanks

We want to thank ThinkR for allowing us time to write this book, and for always being supportive during the project.

Colin wants to personally thank Christophe Dervieux[13] for all his help with this book and any other projects, and for being such an awesome person. He also wants to thank Eric Nantz[14] for taking the time to write the Foreword, for being one of the first {golem} adopters, and for always being enthusiastic about all things gravitating around the golemverse. A big thanks also to David Granjon[15] for his precious feedback on this book.

Contributors to this book

We want to thank all the people that have contributed to this book, either via Pull Requests, Issues on the book's GitHub repository, or via direct conversation.

@aicesan[16], @allaway[17], @andreferraribr[18], @ardeeshany[19], @aronolof[20], @BenjaminLouis[21], @bstrain71[22], @cedricbriandgithub[23], @coreyyanofsky[24], @dA505819[25], @dan-reznik[26], @davidpb90[27], @denrou[28], @Dschaykib[29],

---

[12]https://github.com/ThinkR-open/building-shiny-apps-workflow/issues
[13]https://twitter.com/chrisderv
[14]https://r-podcast.org/
[15]https://divadnojnarg.github.io/
[16]https://github.com/aicesan
[17]https://github.com/allaway
[18]https://github.com/andreferraribr
[19]https://github.com/ardeeshany
[20]https://github.com/aronolof
[21]https://github.com/BenjaminLouis
[22]https://github.com/bstrain71
[23]https://github.com/cedricbriandgithub
[24]https://github.com/coreyyanofsky
[25]https://github.com/dA505819
[26]https://github.com/dan-reznik
[27]https://github.com/davidpb90
[28]https://github.com/denrou
[29]https://github.com/Dschaykib

@ehesch[30], @eroten[31], @espinielli[32], @etiennebacher[33], @fBedecarrats[34], @flrd[35], @frankschmitt[36], @FrieseWoudloper[37], @giocomai[38], @gowachin[39], @graue70[40], @Guillaume-Lombardo[41], @gwd999[42], @hadley[43], @hsm207[44], @jamespooley[45], @jcrodriguez1989[46], @jonmcalder[47], @josbop[48], @jp-marindiaz[49], @jtelleriar[50], @julianstanley[51], @kdaily[52], @maelle[53], @mayeulk[54], @naomistrandberg[55], @psychometrician[56], @rainiercito[57], @ronniemo[58], @rpodcast[59], @sowla[60], @tellyshia[61], @ukyouhanDVSA[62], @xari[63], and @xvrdm[64].

---

[30] https://github.com/ehesch
[31] https://github.com/eroten
[32] https://github.com/espinielli
[33] https://github.com/etiennebacher
[34] https://github.com/fBedecarrats
[35] https://github.com/flrd
[36] https://github.com/frankschmitt
[37] https://github.com/FrieseWoudloper
[38] https://github.com/giocomai
[39] https://github.com/gowachin
[40] https://github.com/graue70
[41] https://github.com/Guillaume-Lombardo
[42] https://github.com/gwd999
[43] https://github.com/hadley
[44] https://github.com/hsm207
[45] https://github.com/jamespooley
[46] https://github.com/jcrodriguez1989
[47] https://github.com/jonmcalder
[48] https://github.com/josbop
[49] https://github.com/jpmarindiaz
[50] https://github.com/jtelleriar
[51] https://github.com/julianstanley
[52] https://github.com/kdaily
[53] https://github.com/maelle
[54] https://github.com/mayeulk
[55] https://github.com/naomistrandberg
[56] https://github.com/psychometrician
[57] https://github.com/rainiercito
[58] https://github.com/ronniemo
[59] https://github.com/rpodcast
[60] https://github.com/sowla
[61] https://github.com/tellyshia
[62] https://github.com/ukyouhanDVSA
[63] https://github.com/xari
[64] https://github.com/xvrdm

Foreword

As a long-time R user (since Version 2.0.0 back in 2004), I have seen more than a few "game-changing" advancements which transformed my entire workflow and opened the doors to new possibilities I never imagined. One of those came in late 2012 when RStudio released {shiny} to the R community. I was absolutely floored by the very notion that I could create not just a web interface, but a dynamic web interface, all through R code! To give a little perspective, the only web interfaces I had built before Shiny were very utilitarian PHP-powered sites with a MySQL database back-end to summarize local state parks data near my graduate school's location, and let's just say those projects would not win any awards for web design!

I certainly experienced the longtime adage of learning the hard way as I began to create Shiny apps at my day job and for personal projects. Over the first year or so of my time with Shiny, I created small apps that revealed the potential it could bring, and it is still amazing that I somehow stitched those together without fully understanding the nuances of reactivity, optimal UI designs, and other software-development principles that a classically trained statistician and Linux enthusiast never knew about! Things began to click in my mind bit by bit (especially after attending the first (and only) Shiny Developer Conference), and I found myself with the task of creating not just simple prototypes, but large-scale software products meant for **production** use. Oh my, what have I gotten myself into?

While being a frequent visitor to the Shiny mailing list and the helpful shiny tag on Stack Overflow, I felt a serious lack of resources addressing the optimal techniques, best practices, and practical advice of taking my Shiny apps to production. And then, one of the most transformative events in my R usage occurred. During the 2019 rstudio::conf, I was checking out the excellent poster session and found the Building Big Shiny Apps[65] poster presented by Colin Fay. I had known Colin as a fellow curator for the RWeekly[66] project and knew he had done some work with Shiny, but during his walkthrough I always had this loud voice in my head saying "Hey, Colin knows exactly what I've been thinking about!" This was the first time I saw the important challenges any Shiny app developer in this space will undoubtedly encounter stated in language I could relate to, even with me being new to the software

---

[65]https://thinkr-open.github.io/rstudioconf2019
[66]https://rweekly.org/

development mindset. Needless to say, I had tremendous fun talking Shiny and all things R with Colin and Vincent Guyader at the conferences, trying to soak up all of their insights and advice every chance I could.

Colin and I both agreed in our Shiny Developer Series episode[67] that creating resources for this audience was an important step in the evolution of sharing best practices with Shiny. Fast forward to today, and you are now reading a tremendous resource aimed squarely at the R users in our world who have embarked on creating production-level applications. *Engineering Production-Grade Shiny Apps* contains an excellent blend of both Shiny-specific topics (many of which have not been addressed in previous books about Shiny) and practical advice from software development that fit in nicely with Shiny apps. You will find many nuggets of wisdom sprinkled throughout these chapters. It's very hard to pick favorites, but certainly one that felt like a moment of enlightenment was the concept of building triggers and watchers to define your own patterns of object invalidation. Now I use that technique in every app I create! Of course, one of the key pillars holding the foundation of this book is the {golem}[68] package, and I have found that the time I invested to learn the ins and outs of creating applications with {golem} has paid off significantly for creating my complex applications, especially with multi-person development teams. As I was finishing my writing of this Foreword, my four-year-old son asked me, "Why does {golem} create nice things, Daddy?" Well, this book is easily the best way to explain that answer! I hope reading *Engineering Production-Grade Shiny Apps* helps you on your journey to creating large Shiny applications!

Eric Nantz - Host of the R-Podcast and the Shiny Developer Series

---

[67]https://shinydevseries.com/post/episode-2-golem
[68]https://thinkr-open.github.io/golem

Application presentation

This book uses a series of applications as examples.

{hexmake}

{hexmake} is an application that has been designed to build hex logos. It was built by Colin, and it serves two main purposes: it helps the creation of a logo, but mainly it serves as an example of some complex features you can use inside a {shiny} application (image manipulation, custom CSS, linking to an external database, save and restore, etc.).

Figure 1 is a screenshot of this application.

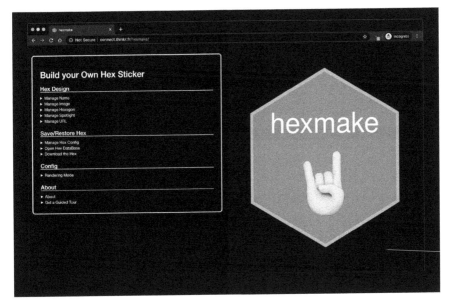

FIGURE 1: The {hexmake} application.

The app is available at engineering-shiny.org/hexmake/[69].

The code is available at github.com/ColinFay/hexmake[70].

{tidytuesday201942}

{tidytuesday201942} is an application using the dataset from week 42 of tidytuesday 2019. It was built by Colin, and it serves as an example of an app built from scratch using bootstrap 4.

Figure 2 is a screenshot of this application.

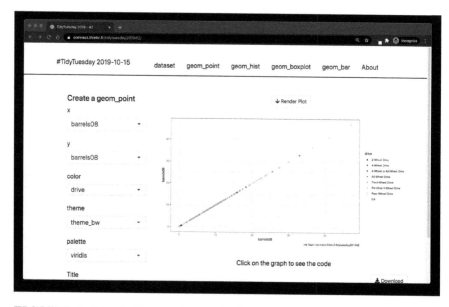

FIGURE 2: The {tidytuesday201942} application.

The app is available at engineering-shiny.org/tidytuesday201942/[71].

The code is available at github.com/ColinFay/tidytuesday201942[72].

---

[69]https://engineering-shiny.org/hexmake/
[70]https://github.com/ColinFay/hexmake
[71]https://engineering-shiny.org/tidytuesday201942/
[72]https://github.com/ColinFay/tidytuesday201942

{minifying}

{minifying} is an application to minify CSS, JavaScript, HTML, and JSON files. It was built built by Colin as a use case for the workflow of this book. You will find the details of how this app was constructed in the Appendix, "*Use case: Building an App, from Start to Finish*".

Figure 3 is a screenshot of this application.

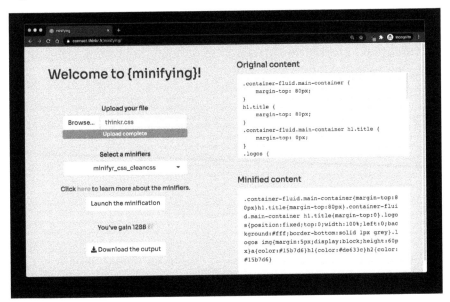

FIGURE 3: The {tidytuesday201942} application.

The app is available at engineering-shiny.org/minifying/[73]. The code is available at github.com/ColinFay/minifying[74].

Other applications

- {shinipsumdemo} is an application built by Cervan as an example for {shinipsum}, available at engineering-shiny.org/shinipsumdemo/[75].

---

[73]https://engineering-shiny.org/minifying/
[74]https://github.com/ColinFay/minifying
[75]https://engineering-shiny.org/shinipsumdemo/

- {golemhtmltemplate} is an application built by Colin as an example for {shinipsum} and shiny::htmlTemplate(), available at engineering-shiny.org/golemhtmltemplate/[76].

- {databasedemo} is an application built by Cervan using an external database, available at engineering-shiny.org/databasedemo/[77].

- {grayscale} is an application built by Cervan using an external html template, available at engineering-shiny.org/grayscale/[78].

- {bs4dashdemo} is an application built by Cervan with the {bs4dash} package, available at engineering-shiny.org/bs4dashdemo/[79].

- {shinyfuture} is an application built by Cervan as an example of using {promises} and {future} inside a {shiny} app, available at engineering-shiny.org/shinyfuture/[80].

---

[76]https://engineering-shiny.org/golemhtmltemplate/
[77]https://engineering-shiny.org/databasedemo/
[78]https://engineering-shiny.org/grayscale/
[79]https://engineering-shiny.org/bs4dashdemo/
[80]https://engineering-shiny.org/shinyfuture/

Part I

Building Successful {shiny} Apps

1

About Successful {shiny} Apps

1.1 A (very) short introduction to {shiny}

If you are reading this page, chances are you already know what a {shiny} application (sometimes shortened as "{shiny} app") is—**a web application that communicates with R, built in R, and working with R**. The beauty of {shiny} (Chang et al., 2021a) is that it makes it easy for someone already familiar with R to create a small app in a matter of hours. With small and minimal {shiny} apps, no knowledge of HTML (HyperText Markup Language), CSS(Cascading Style Sheets) or JavaScript is required, and you do not have to think about technical elements that usually come with web applications—for example, you do not have to think about the port the application is served on: {shiny} picks one for you.[1] Same goes for serving external dependencies: the application comes with its set of CSS and JavaScript dependencies that a common {shiny} developer does not need to worry about. And that is probably one of the main reasons why this package has become so successful over the years—**with very little training, you can rapidly create a proof-of-concept (PoC) for a data product, showcase an algorithm, or present your results in an elegant and accessible user interfaces.**

The first version of {shiny} was published in 2012. Since then, it has been one of the top projects of the RStudio team. At the time of writing these lines (April 2020), there are more than 4700 commits in the master branch of the GitHub repository, made by 46 contributors. It is now downloaded around 400K times a month, according to cranlogs[2], and has 922 reverse dependencies (i.e. packages that depend on it), according to `revdep("shiny")` from {devtools} (Wickham et al., 2021d).

If you are very new to {shiny}, this book might feel a little bit overwhelming: we will be discussing some advanced {shiny} and software engineering methods, best practices and structural ideas for sending {shiny} applications to production. This book relies on the assumption that you already know

---

[1] Of course you can specify one if you need to, but by default the package picks one.
[2] https://cranlogs.r-pkg.org/badges/shiny

how to build basic {shiny} applications, and that you want to push your {shiny} skills to the next level: in other words, you are ready to move from the Proof of Concept to the production-grade application. If you are very new to {shiny}, we suggest you start with the *Mastering Shiny* book[3] before reading the present book.

Ready to start engineering production-grade {shiny} apps?

1.2   What is a complex {shiny} application?

> One of the unfortunate things about reality is that it often poses complex problems that demand complex solutions.
>
> *The Art of Unix Programming* (Raymond, 2003)

1.2.1   Reaching the cliff of complexity

Building a {shiny} application seems quite straightforward when it comes to small prototypes or proof of concepts: after a few hours of practice and documentation reading, most R developers can have a small working application. But things change when your application reaches "the cliff of complexity",[4] i.e. that moment when the application reaches a state when it can be qualified as "complex".

But what do we mean by complexity? Getting a clear definition is not an easy task [5] as it very much depends on who is concerned and who you are talking to. But a good definition can be found in *The DevOps Handbook* (Kim, 2016): "One of the defining characteristics of a complex system is that it **defies any single person's ability to see the system as a whole and understand how all the pieces fit together.** Complex systems typically have a high degree of interconnectedness of tightly coupled components, and system-level

---

[3]https://mastering-shiny.org/

[4]We borrow this term from Charity Major, as heard in *Test in Production with Charity Majors* CoRecursive #019, *Aug 31, 2018*.

[5]Ironic, right?

behavior cannot be explained merely in terms of the behavior of the system components." (Our bold.)

Or as noted in *Refactoring at Scale* (Lemaire, 2020), "It becomes nearly impossible to reason about the effect a change might have when applied uniformly across a sprawling, complex system. Many tools exist to identify code smells or automatically detect improvements within subsections of code, but we are largely unable to automate human reasoning about how to restructure large applications, in codebases that are growing at an increasingly rapid pace."

Building on top of these quotes, let's try to come up with a definition that will serve us in the context of engineering {shiny} applications.

When building software, we can think of complexity from two points of view: the complexity as it is seen by the developer, and the complexity as it is seen by the customer/end user.[6]

- For the code, **bugs are harder to anticipate**: it is hard to think about all the different paths the software can follow and difficult to identify bugs because they are deeply nested in the numerous routines the app is doing. It is also hard to think about what the state of your app is at a given moment because of the numerous inputs and outputs your app contains.

- **From the user perspective, the more complex an app is, the steeper the learning curve.** Indeed, the user will have to invest more time learning how the app works, and will be even more disappointed if ever they realize this time has been a waste.

Let's dive a little bit more into these **two types of complexity**.

A. Developer complexity

An app is to be considered complex when it is so large in terms of size and functionality that it makes it impossible to reason about it at once, and **developers must rely on tools and methods to understand and handle this complexity**: for example, when it comes to {shiny}, you will rely on tools like the {golem} (Fay et al., 2021) framework, introduced throughout this book, to handle some implementation, development and deployment complexity. This book will introduce a clear methodology that comes with a series of conventions, which are crucial when it comes to building and maintaining complex systems: by imposing a formalized structure for software, it enhances its readability, lowers the learning curve for newcomers, and reduces the risk of errors inherent in repetitive tasks.

---

[6]From *The Art of Unix Programming*, "Chapter 13: Speaking of Complexity" (Raymond, 2003).

This type of complexity is called *implementation complexity*. One of the goals of this book is to present a methodology and toolkit that will help you reduce this form of complexity.

B. Customer and user complexity

Customers and end users see complexity as *interface complexity*.

Interface complexity can be driven by a lot of elements, for example, the probability of making an error while using the app, the difficulty in understanding the logical progression in the app, the presence of unfamiliar behavior or terms, visual distractions, etc. This book will also bring you strategy to help you cope with the need for simplification when it comes to designing an interface.

1.2.2 Balancing complexities

There is an inherent tension between these two sources of complexity, as designing an app means finding a good balance between implementation and interface complexity. **Lowering one source of complexity usually means increasing the other, and managing an application project means knowing where to draw the line.** This usually requires restraining yourself from implementing too many features, and still creating an application that is easy to use, and that fits the requirements you have received.

For example, there is something common in {shiny} applications: what we can call the "too much reactivity pattern". In some cases, developers try to make everything reactive: *e.g.*, three sliders and a drop-down input, all updating a single plot. This behavior lowers the interface complexity: users do not have to really think about what they are doing, they move sliders, change the inputs, and boom! the plot updates. But this kind of pattern can make the application perform too much computation, for example, because users rarely go to the value they need on their first try: they usually miss the value they actually want to select.

One solution can be to delay reactivity or to cache things so that R computes fewer things. But that comes with a cost: handling delayed reactivity and caching elements increase implementation complexity. Another solution is to add an "update plot" button, which updates the plot only when the user clicks on it. This pattern makes it easier to control reactivity from the implementation side. But this can make the interface a little bit more complex for the users, who have to perform another action, on top of changing their inputs. We will argue in this book that not enough reactivity is better than too much reactivity, as the latter increases computation time, and relies on the assumption that the user makes the right action on the first try.

Another good example is {shiny}'s `dateRangeInput()` function. This function requires the user to choose a start date and an end date. However, the function allows the user to choose a start date which is before the end (that is the behavior of the JavaScript plugin used in {shiny} to create this input). But allowing this behavior leads to bugs, notably in a context of full reactivity. Handling this special case is completely doable: with a little bit of craft, you can watch what the user inputs and throw an error if the start is after the end.[7] On one hand, that solution increases implementation complexity, while on the other hand, allowing this naive behavior requires the user to think carefully about what they are selecting, thus increasing the interface complexity.

What should we do? It's up to you: deciding where to draw the line between interface and implementation complexity very much depends on the project, but that is something that you should keep in mind throughout the project's life.

1.2.3 Assessing code complexity

On the developer side, you will want to **reduce code complexity so that everybody involved in the coding process is able to create a mental model of how the application works.** On the user side, you will want to **reduce interface complexity so that everybody comes out of using your application with a good user experience.**

Reducing complexity first comes with being able to identify its potential sources, be it in your application codebase or in the specifications describing how the application should work. Finding these sources of complexity is not an easy task, as it requires some programming knowledge to identify bottlenecks, basic UX (User Experience) skills to implement a good interface, and of course a project management methodology to structure the whole life of your application.

All these points will be addressed in this book. But before that, let's dive into code complexity.

A. Codebase size

The total number of lines of code, and the number of files, can be good clue of potential complexity, but only if used as an order of magnitude (for example, a 10,000-line codebase is potentially more complex than a 100-line codebase), but should not be relied on if used strictly, even more if you try to reduce the number of lines by sacrificing code readability.

R is very permissive when it comes to indentation and line breaks, and, unlike

---

[7]See shiny/issues/2043#issuecomment-525640738[8] for an example.

JavaScript or CSS, it is generally not minified.[9] In R, the number of lines of code depends on your coding style and the packages you are using. For example, the {tidyverse} (Wickham et al., 2019) style guide encourages the use of %>% (called "pipe"), with one function by line, producing more lines in the end code: "%>% should always have a space before it, and should usually be followed by a new line" (tidyverse style guide[10]). So you can expect a "tidyverse-centric" package to contain more lines of code, yet the pipe itself has been thought of as a tool to lower code complexity by enhancing its readability.[11]

For example, the two following pieces of code do the same thing. Both have a different number of lines, and a different level of reading complexity.

```r
library(dplyr, warn.conflicts = FALSE)
# With pipe
iris %>%
  group_by(Species) %>%
  summarize(mean_sl = mean(Sepal.Length))
# Without the pipe
summarize(group_by(iris, Species), mean_sl = mean(Sepal.Length))
```

Also, there is no limit in the way you can indent your code.

```r
# Putting one symbol by line
iris[
  1
  :
    5,
  "Species"
]
```

Six lines of code for something that could also be written in one line.

```r
# Same code but everything is on the same line
iris[1:5, "Species"]
```

---

[9]The minification process is the process of removing all blank characters and putting everything on one line so that the file in the output is much smaller.

[10]https://style.tidyverse.org/pipes.html

[11]Note though that some users find using the pipe more complex.

In other words, using this kind of writing style can make the codebase larger in term of lines, without really adding complexity to the general program.

Another drawback of this metric is that it focuses on numbers instead of readability, and in the long run, yes, readability matters. As noted in *The Art of Unix Programming*, "Pressure to keep the codebase size down by using extremely dense and complicated implementation techniques can cause a cascade of implementation complexity in the system, leading to an un-debuggable mess" (Raymond, 2003).

Still, this metric can be useful to reinforce what you have learned from other metrics. It is rather unlikely that you will find this "extreme" coding style we showed above, and even if it might not make sense to compare two codebases that just differ by 1% or 2 % of lines of code, it is very likely that a codebase which is ten, one hundred, one thousand times larger is a more complex software.

Another good metric related to code complexity is the number of files in the project: R developers tend to split their functions into several files, so the more files you will find in a project, the larger the codebase is. And numerous files can also be a sign of maintenance complexity, as it may be harder to reason about an app logic that is split into several files than about something that fits into one linear code inside one file.[12] On the other hand, one big 10,000-line file which is standing alone in the project is not a good sign either.

If you want to use the number-of-lines metric, you can do it from R with the {cloc} (Rudis and Danial, 2020) package, available at `https://github.com/hrbrmstr/cloc`.

```r
# Install {cloc} from GitHub
remotes::install_github("hrbrmstr/cloc")
```

For example, let's compare a rather big package ({shiny}) with a small one ({attempt} (Fay, 2020)):

```r
library(cloc)
# Using dplyr to manipulate the results
library(dplyr, warn.conflicts = FALSE)

# Computing the number of lines of code
```

---

[12]To handle the complexity of splitting into files, you can set filenames to follow the structure of the project. This pattern is developed in another part of this book, where we explain the conventions used in {golem}.

```r
# for various CRAN packages
shiny_cloc <- cloc_cran(
  "shiny",
  .progress = FALSE,
  repos = "http://cran.irsn.fr/"
)
attempt_cloc <- cloc_cran(
  "attempt",
  .progress = FALSE,
  repos = "http://cran.irsn.fr/"
)

clocs <- bind_rows(
  shiny_cloc,
  attempt_cloc
)

# Summarizing the number of line of code inside each package
clocs %>%
  group_by(pkg) %>%
  summarise(
    loc = sum(loc)
  )
```

```
# A tibble: 2 x 2
  pkg          loc
  <chr>      <int>
1 attempt     6486
2 shiny     175376
```

```r
# Summarizing the number of files inside each package
clocs %>%
  group_by(pkg) %>%
  summarise(
    files = sum(file_count)
  )
```

```
# A tibble: 2 x 2
  pkg        files
  <chr>      <int>
1 attempt       64
2 shiny        736
```

Here, with these two metrics, we can safely assume that {shiny} is a more complex package than {attempt}. If you want to compute the same prefix for a local package/repository, the `cloc_pkg()` function can be used. For example, here is how to compute the cloc metric for the {hexmake} application:

```
# Calling the function on the {hexmake}
# application Git repository
hexmake_cloc <- cloc_git(
  "https://github.com/ColinFay/hexmake"
)
hexmake_cloc
```

```
# A tibble: 10 x 10
   source   language   file_count file_count_pct   loc
   <chr>    <chr>          <int>            <dbl> <int>
 1 hexmake  JSON               1           0.0102  3844
 2 hexmake  R                 34           0.347   2345
 3 hexmake  Markdown           5           0.0510    95
 4 hexmake  CSS                1           0.0102    76
 5 hexmake  Dockerfile         2           0.0204    45
 6 hexmake  JavaScript         2           0.0204    31
 7 hexmake  Rmd                2           0.0204    18
 8 hexmake  HTML               1           0.0102    14
 9 hexmake  YAML               1           0.0102     8
10 hexmake  SUM               49           0.5     6476
# ... with 5 more variables: loc_pct <dbl>,
#    blank_lines <int>, blank_line_pct <dbl>,
#    comment_lines <int>, comment_line_pct <dbl>
```

One thing that this package also allows is counting the number of lines of commented code: it's usually a good sign to see that a package has comments in its code-base, as it will allow to work more safely in the future, provided that this metric doesn't reveal that large portions of the application are "commented code" (as opposed to "code comments"). For example, here we can see that {hemake} has 2345 lines of R code, which come with 669 lines of code comments.

B. Cyclomatic complexity

Cyclomatic complexity is a software engineering measure which **allows us to define the number of different linear paths a piece of code can take**. The higher the number of paths, the harder it can be to have a clear mental model of this function.

Cyclomatic complexity is computed based on a control-flow graph [13] representation of an algorithm, as can be seen on Figure 1.1. For example, here is a
simple control flow for an `ifelse` statement *(The following paragraph details
the algorithm implementation, feel free to skip it if you are not interested in
the implementation details).*

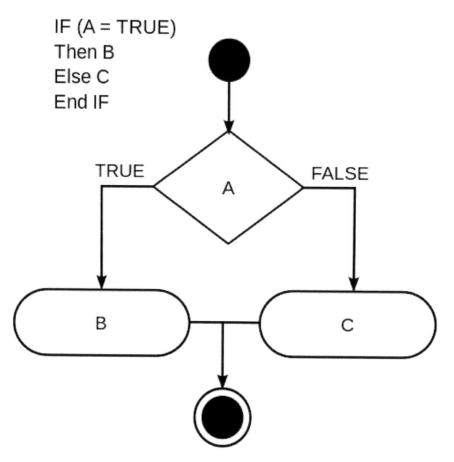

FIGURE 1.1: Control-flow graph representation of an algorithm.

The complexity number is then computed by taking the number of nodes,
subtracting the number of edges, and adding twice the number of connected
components of this graph. The algorithm is then $M = E - N + 2P$, where
M is the measure, E the number of edges, N the number of nodes and $2P$
is twice the number of connected components. We will not go deep into this
topic, as there are a lot things going on in this computation and you can find

---

[13] A control flow graph is a graph representing all the possible paths a piece of code can
take while it is executed.

much documentation about this online. Please refer to the bibliography for further readings about the theory behind this measurement.

In R, the cyclomatic complexity can be computed using the {cyclocomp} (Csardi, 2016) package. You can get it from `CRAN` with:

```r
# Install the {cyclocomp} package
install.packages("cyclocomp")
```

The {cyclocomp} package comes with three main functions: `cyclocomp()`, `cyclocomp_package()`, and `cyclocomp_package_dir()`. While developing your application, the one you will be interested in is `cyclocomp_package_dir()`: building successful shiny apps with the {golem} framework means you will be building your app as a package (we will get back on that later).

Here is, for example, the cyclomatic complexity of the default golem template (assuming it is located in a **golex/** subdirectory):

```r
# Launch the {cyclocomp} package, and compute the
# cyclomatic complexity of "golex",
# A blank {golem} project with one module skeleton
library(cyclocomp)
cyclocomp_package_dir("golex") %>%
  head()
```

```
                       name cyclocomp
1                 app_server         1
2                     app_ui         1
3 golem_add_external_resources         1
4                    run_app         1
```

And the one from another small application:

```r
# Same metric, but for the application
# {tidytuesday201942}, available at
# https://engineering-shiny.org/tidytuesday201942.html
cyclocomp_package("tidytuesday201942") %>%
  head()
```

```
              name cyclocomp
24       mod_dataviz_ui         8
23   mod_dataviz_server         7
35                   rv         6
14              display         4
39            undisplay         4
37 tagRemoveAttributes         3
```

And, finally, the same metric for {shiny}:

```
# Computing this metric for the {shiny} package
cyclocomp_package("shiny") %>%
  head()
```

```
                     name cyclocomp
494                untar2        75
115           diagnoseCode        54
389                runApp        50
150 find_panel_info_non_api      37
371           renderTable        37
102          dataTablesJSON        34
```

And, bonus, this `cyclocomp_package()` function can also be used to retrieve the number of functions inside the package.

As The Clash said, "What are we gonna do now?" You might have heard this saying: "if you copy and paste a piece of code twice, you should write a function", so you might be tempted to do that. Indeed, splitting code into smaller pieces lowers the local cyclomatic complexity, as smaller functions have lower cyclomatic complexity. But that is just at a local level, and it can be a suboptimal option: having a very large number of functions calling each other can make it harder to navigate through the codebase.

In the end of the day, splitting into smaller functions is not a magic solution because:

- the global complexity of the app is not lowered by splitting things into pieces (just local complexity) and
- tThe deeper the call stack, the harder it can be to debug.

C. Other measures for code complexity

Complexity can come from other sources: **insufficient code coverage, dependencies that break the implementation, relying on old packages**, or a lot of other things.

We can use the {packageMetrics2} (Csardi, 2021) package to get some of these metrics: for example, the number of dependencies, the code coverage, the number of releases and the date of the last one, etc., and the number of lines of code and the cyclomatic complexity.

At the time of writing these lines, the package is not on CRAN and can be installed using the following line of code:

This package can now be used to assess the dependencies we use in our application. To do that, let's create a small function that computes this metric and returns a tibble:

```
library()
# A function to turn the output of the metrics into a data.frame
frame_metric <- function(pkg){
  metrics <- package_metrics(pkg)
  tibble::tibble(
    n = names(metrics),
    val = metrics,
    expl = list_package_metrics()[names(metrics)]
  )
}
```

And run the metric for {golem},

```
# Using this function with{golem}
frame_metric("golem")
```

```
# A tibble: 22 x 3
   n     val                  expl
   <chr> <chr>                <chr>
 1 ARR   4                    Number of times = is used f~
 2 ATC   <NA>                 Author Test Coverage
 3 DWL   82823                Number of Downloads
 4 DEP   33                   Num of Dependencies
 5 DPD   12                   Number of Reverse-Dependenc~
 6 CCP   2.72727272727273     Cyclomatic Complexity
 7 FLE   4.23440285204991     Average number of code line~
 8 FRE   2019-08-05T14:50~    Date of First Release
 9 LIB   0                    Number of library and requi~
10 LLE   109                  Number of code lines longer~
# ... with 12 more rows
```

And {shiny}

```
# Using this function with {shiny}
frame_metric("shiny")
```

```
# A tibble: 22 x 3
     n     val                expl
     <chr> <chr>              <chr>
  1 ARR   14                 Number of times = is used f~
  2 ATC   46.3623848353278   Author Test Coverage
  3 DWL   18122047           Number of Downloads
  4 DEP   35                 Num of Dependencies
  5 DPD   933                Number of Reverse-Dependenc~
  6 CCP   3.80062794348509   Cyclomatic Complexity
  7 FLE   2.41493883924699   Average number of code line~
  8 FRE   2012-12-01T07:16~  Date of First Release
  9 LIB   4                  Number of library and requi~
 10 LLE   738                Number of code lines longer~
# ... with 12 more rows
```

If you are building your {shiny} application with {golem}, you can use the
DESCRIPTION file, which contains the list of dependencies, as a starting point
to explore these metrics for your dependencies, for example, using {desc}
(Csárdi et al., 2021) or {attachment} (Rochette and Guyader, 2021):

```
# Get the dependencies from the DESCRIPTION file.
# You can use one of these two functions to list
# the dependencies of your package,
# and compute the metric for each dep
desc::desc_get_deps("golex/DESCRIPTION")
```

```
    type package version
1 Imports  config       *
2 Imports   golem       *
3 Imports   shiny       *
```

```
# See also
attachment::att_from_description("golex/DESCRIPTION")
```

```
[1] "config" "golem"  "shiny"
```

Some important metrics to watch there are as follow:

- Test coverage: the more the better, as a large code coverage should imply that bugs are more easily caught.
- The number of downloads: a largely downloaded package will likely be less prone to bug, as it will be used by a large user base.
- Number of dependencies: the more a package has dependencies, the more likely it is that at some point it time, something in the dependency graph will break.
- Dates of first publish on CRAN, last publish, and updates: a package actively maintained is a good sign.[14]

D. Complexity assessment checklist

To sum up, here is a quick checklist of things to check to assess the complexity of your application:

☐ Running the metrics from {cloc}, to get an idea of the number of files, their diversity in terms of extensions (for example {hexmake} also has JSON, JavaScript, and YAML files), and the ratio of comments for the code. Remember that having only one big R file is a red flag, and so is having zero code comments.

☐ Assess the cyclomatic complexity of the package containing your application. Remember that the more a function scores on this metric, the more complex it will be to debug it.

☐ Check the package common metrics using {packageMetrics2}, notably for the dependencies you are including in your package. Metrics to look for are test coverage, number of downloads, number of dependencies, and date of first release and last release.

---

[14]Even if this is not an absolute rule, some packages haven't been updated for a long time but are still completely reliable.

1.2.4   Production-grade software engineering

Complexity is still frowned upon by a lot of developers, notably because it has been seen as something to avoid according to the Unix philosophy. But there are dozens of reasons why an app can become complex: for example, the question your app is answering is quite complex and involves a lot of computation and routines. The resulting app is rather ambitious and implements a lot of features, etc. There is a chance that if you are reading this page, you are working or are planning to work on a complex {shiny} app. And this is not necessarily a bad thing! {shiny} apps can definitely be used to implement production-grade [15] software, but production-grade software implies production-grade software engineering. To make your project a success, you need to use tools that reduce the complexity of your app and ensure that your app is resilient to aging.

In other words, production-grade {shiny} apps require working with a software engineering mindset, which is not always an easy task in the R world: many R developers have learned this language as a tool for doing data analysis, building model, and making statistics; not really as a tool for building software.

The use of R has evolved since its initial version released in 1995, and using this programming language as a tool to build software for production is still a challenge, even 26 years after its first release. And still today, for a lot of R users, the software is still used as an "experimentation tool", where production quality is one of the least concerns. But the rise of {shiny} (among other packages) has drastically changed the potential of R as a language for production software engineering: its ease of use is also one of the reasons why the language is now used outside academia, in more "traditional" software engineering teams.

This changing context requires different mindsets, skills, and tools.

With {shiny}, as we said before, it is quite easy to prototype a simple app, without any "hardcore" software engineering skills. And when we are happy with our little proof of concept, we are tempted to add something new. And another. And another. And **without any structured methodology, we are almost certain to reach the cliff of complexity very soon and end up with a codebase that is hardly (if ever) ready to be refactored to be sent to production.**

The good news is that building a complex app with R (or with any other language) is not an impossible task. But this requires planning, rigor, and correct engineering. This is what this book is about: how to organize your

---

[15]By production-grade, we mean a software that can be used in a context where people use it for doing their job, and where failures or bugs have real-life consequences.

{shiny} app in a way that is time and code efficient, and how to use correct engineering to make your app a success.

1.3 What is a successful {shiny} app?

Defining what "successful" means is not an easy task, but we can extract some common patterns when it comes to applications that would be considered successful.

1.3.1 It exists

First of all, an app is successful if it was delivered. In other words, **the developer team was able to move from specification to implementation to testing to delivering.** This is a very engineering-oriented definition of success, but it is a pragmatic one: an app that never reaches the state of usability is not a successful app, and something along the way has blocked the process of finishing the software.

This condition implies a lot of things, but mostly it implies that the team members were able to organize themselves in an efficient way, so that they were able to work together in making the project a success. Anybody that has already worked on a codebase as a team knows it is not an easy task.

1.3.2 It is accurate

The project is a success if the application was delivered, and if **it answers the question it is supposed to answer, or serves the purpose it is supposed to serve.** Delivering is not the only thing to keep in mind: you can deliver a working app but it might not work the way it is supposed to work.

Just as before, accuracy means that between the moment the idea appears in someone's mind and the moment the app is actually ready to be used, everybody was able to work together toward a common goal, and now that this goal is reached, we are also certain that the answers we get from the application are accurate, and that users can rely on the application to make decisions.

1.3.3   It is usable

Being usable means that the app was delivered, it serves the purpose, and it is user-friendly.

Unless you are just coding for the joy of coding, there will always be one or more end users. And **if these people cannot use the application because it is too hard to use, too hard to understand, because it is too slow or there is no inherent logic in how the user experience is designed, then it is inappropriate to call the app a success.**

1.3.4   It is immortal

Of course, "immortal" is a little bit far-fetched, but when designing the application, you should aim for robustness through the years, by engineering a (theoretically) immortal application.

Planning for the future is a very important component of a successful {shiny} app project. Once the app is out, it is successful if it can **exist in the long run, with all the hazards that this implies**: new package versions that could potentially break the codebase, sudden calls for the implementation of new features in the global interface, changing key features of the UI (User Interface) or the back-end, not to mention passing the codebase along to someone who has not worked on the first version, and who is now in charge of developing the next version.[16] And this, again, is hard to do without effective planning and efficient engineering.

---

[16] In fact, this new person might simply be you, a month from now. And *"You'll be there in the future too, maintaining code you may have half forgotten under the press of more recent projects. When you design for the future, the sanity you save may be your own.* (Raymond, 2003).

2

2.1 Working with a "long-term" mindset

Rome ne fut pas faite toute en un jour.

French proverb

2.1.1 Prepare for success

Whatever your ambitions for your {shiny} application, you should take time today to set robust foundations that will save a lot of time in the future.

A common thing you will hear about {shiny} is that it is a good prototyping tool. This cannot be denied. Building a Proof of Concept (PoC) for an app is relatively straightforward if you compare to what is needed to build applications in other languages. With {shiny}, you can build an "it works on my machine" web application in a couple of hours, and show it to your team, your boss, your investors. Thanks to the way {shiny} was designed, you do not have to care about websockets, ports, HTML (HyperText Markup Language), JavaScript libraries, and all the things that are elegantly bundled into {shiny}.

Hence, you can have a quick, hacky application that will work on your machine, and very rapidly. But that is not the way you should start. Indeed, starting with hacky foundations will lead to two possibilities:

- You will have to **rewrite everything from scratch** to have a robust application for production.
- If you do not want to rewrite all the code, you will **get stuck with a legacy codebase** for the application, built on top of hacky functions, and sent to production using hacky solutions.

21

Either way, that is a **heavy technical debt**.

{shiny} is a good tool for prototyping, but there is no harm in starting your application on solid ground, even for a prototype: **the sooner you start with a robust framework the better, and the longer you wait the harder it gets to convert your application to a production-ready one**. The larger the codebase, the harder it is to untangle everything and make it work.

In this book, we will present a framework called {golem}, which is a toolbox for building production-grade {shiny} applications. Even if {golem} is focused on production, there is no reason not to use it for your proof of concepts: starting a new {golem} project is relatively straightforward, and even if you do not use the advanced features, you can use it for very small apps. The benefit of starting straight inside a {golem} application really outweighs the cost. We hear a lot the question "When should I switch to {golem}?" The answer is simple: do not switch to {golem}, start with it. That way, you are getting ready for complexity, and if, one day, you need to turn this small app into a production app, the foundations are there.

2.1.2   Develop with the KISS principle

The KISS principle states that most systems work best if they are kept simple rather than made complicated; therefore, simplicity should be a key goal in design, and unnecessary complexity should be avoided.

KISS principle, Wikipedia article (https://en.wikipedia.org/wiki/KISS_principle)

The KISS principle, as "Keep It Simple, Stupid", should drive the implementation of features in the application to allow anyone in the future, including original developers, to take over on the development.

The story behind this principle is supposed to be that Kelly Johnson, lead engineer at the Lockheed Skunk Works, gave his workers a set of very common tools and said that every airplane should be repairable with these tools, and these tools only, so that repairing an aircraft should be possible for any average engineer.

This should be a principle to keep in mind when building applications. Indeed, large-scale {shiny} projects can lead to many people working on the codebase, for a long period of time. **A large team means a variety of skills**, with some common ground in {shiny} development, but potentially various levels when it comes to R, web development, or production engineering. When choosing how and what to implement, **try to make a rule to go for the simplest solution,**[1] *i.e.* the one that any common {shiny} developer would be able to understand and maintain. If you go for an exotic solution or a complex technology, be sure that you are doing it for a good reason: unknown or hard-to-grasp technology reduces the chance of finding someone that will be able to maintain that piece of code in the future, and reduce the smoothness of collaboration, as *"Code you can easily comprehend elevates absolutely everyone on your team, no matter their tenure or experience level"* (Lemaire, 2020).

2.2 Working as a team: Tools and structure

Working as a team, whatever the coding project, requires adequate tools, discipline and organization. Complex {shiny} apps usually imply that several people will work on the application. For example, at ThinkR[2], 3 to 4 people usually work in parallel on the same application, but there might be more people involved on larger projects. **The choice of tools and how the team is structured is crucial for a successful application.**

2.2.1 From the tools point of view

A. Version control and test all things

To get informed about a code break during development, you will need to write tests for your app, and use continuous integration (CI) so that you are sure this is automatically detected.[3] When you are working on a complex application, chances are that you will be working on it for a significant period of time, meaning that you will write code, modify it, use it, go back to it after a few weeks, change some other things, and probably break things. **Breaking things is a natural process of software engineering, notably when working on a piece of code during a long period.** Remember the last

---

[1]Which might not be the most "elegant" solution, but production code requires pragmatism.

[2]//rtask.thinkr.fr

[3]Relying on automatic tooling for monitoring the codebase is way safer than relying on developers to do manual checks every time they commit code.

chapter where we explained that complex applications are too large to be understood fully? Adding code that breaks the codebase will happen with complex apps, so the sooner you take measures to solve these changes, the better.

As you cannot prevent code from breaking, you should at least get the tooling to:

- **Be informed that the codebase is broken**: this is the role of tests combined with CI.
- **Be able to identify changes between versions, and potentially, get back in time to a previous codebase**: this is the role of version control.

We will go deeper into testing and version control in chapter 14.

B. Small is beautiful

Building an application with multiple small and independent pieces will lighten your development process and your mental load. The previous chapter introduced the notion of complexity in size, where the app grows so large that it is very hard to have a good grasp of it. **A large codebase implies that the safe way to work is to split the app into pieces.**

Splitting a {shiny} project is made possible by following two methods:

- **Extract your core "non-reactive" functions, which we will also call the "business logic", and include them in external files,** so that you can work on these outside of the app. Working on independent functions to implement features will prevent you from relaunching the whole application every time you need to add something new.

- **Split your app into {shiny} modules,** so that your app can be though of as a tree, making it possible for every developer to concentrate on one node, and only one, instead of having to think about the global infrastructure when implementing features.

Figure 2.1 is, for example, a representation of a {shiny} application with modules and sub-modules. You will not be able to decipher the text inside the node, but the idea is to give you a sense of how a {shiny} application with modulse can be organized and split into smaller pieces that are all related to each other in a tree form.

We will get back to {shiny} modules and how to organize your project in the next chapter.

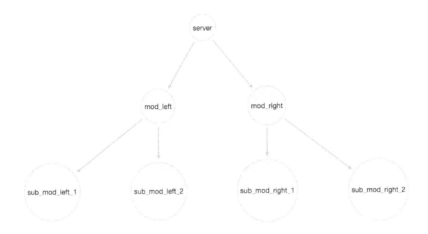

FIGURE 2.1: Representation of a {shiny} application with its modules and sub-modules.

2.2.2 From the team point of view

We recommend that you define two kinds of developers: a unique person (or maybe two) to be in charge of supervising the whole project and developers that will work on specific features. Note that this is how a project should be organized in a perfect world: in practice, a lot of {shiny} projects are managed by one developer who is in charge of managing the project, interacting with the client, and building the whole codebase.

A. A person in charge

The person in charge of the development will have **a complete view of the entire project and manage the team so that all developers implement features that fit together**.

With complex applications, it can be hard to have the complete understanding of what the entire app is doing. Most of the time, it is not necessary for all developers to have this complete picture. By defining one person in charge, this "manager" will have to get the whole picture: what each part of the software is doing, how to make everything work together, avoid development conflicts, and of course check that, at the end of the day, the results returned by the built application are the correct ones.

The project manager will be the one that kicks off the project, and writes the first draft of the application. If you follow this book's workflow, this person will first create a {golem} project, fill in the information, and define the application structure by providing the main modules, and potentially work on the prototyped UI of the app.

Once the skeleton of the app is created, this person in charge will list all the things that have to be done. We strongly suggest that you use `Git` with a graphical interface (GitLab, GitHub, Bitbucket, etc.) as the graphical interface to help you manage the project. These tasks are defined as issues, and will be closed during development. These interfaces can also be used to set continuous integration.

If the team follows a `git flow` (described in Chapter 12), the manager will also be in charge of reviewing and accepting the pull/merge requests to the main `dev` branch if they solve the associated issues.

Do not worry if this sounds like a foreign language to you, we will get back to this method later in this book (Chapter 12).

B. Developers

Developers will focus on small features. If the person in charge has correctly separated the work between developers of the team, they will be focusing on one or more parts of the application, but do not need to know every single bit of what the application is doing. In a perfect world, the application is split in various `{shiny}` modules, one module equals one file, and each member of the team will be assigned to the development of one or more modules.

It is simpler to work in this context where one developer is assigned to one module, although we know that in reality it may be a little more complex, and several members of the team might go back and forth working on a common module. But the person in charge will be there to help make all the pieces fit together.

3

3.1 {shiny} app as a package

In the next chapter you will be introduced to the {golem} (Fay et al., 2021) package, **an opinionated framework for building production-ready {shiny} applications**. This framework will be used a lot throughout this book, and it relies on the idea that every {shiny} application should be built as an R package.

But in a world where {shiny} applications are mostly created as a series of files, why bother with a package?

3.1.1 What is in a production-grade {shiny} app?

You probably haven't realized it yet, but if you have built a significant (in terms of size or complexity of the codebase) {shiny} application, chances are you have been using a package-like structure without knowing it.

Think about your last {shiny} application, which was created as a single-file (app.R) or two-file app (ui.R and server.R). On top of these files, what do you needed to make it **a production-ready application**, and why is a package the perfect infrastructure?

A. It has metadata

First of all, **metadata**. In other words, all the necessary information for something that runs in production: the name of the app, the version number (which is crucial to any serious, production-level project), what the application does, who to contact if something goes wrong, etc.

This is what you will get when using a package DESCRIPTION file.

27

B. It handles dependencies

Second, you need to find a way to **handle the dependencies**. When you want
to push your app into production, you do not want to have this conversation
with the IT team:

- *IT: Hey, I tried to* `source("app.R")` *as you said, but I got an error.*

- *R-dev: What is the error?*

- *IT: It says "could not find package 'shiny'".*

- *R-dev: Ah yes, you need to install {shiny}. Try to run* `install.packages("shiny")`.

- *IT: OK nice. What else?*

- *R-dev: Let me think, try also* `install.packages("DT")`*... good? Now try* `install.packages("ggplot2")`, *and ...*

- *[...]*

- *IT: Ok, now I source the 'app.R', right?*

- *R-dev: Sure!*

- *IT: Ok so it says "could not find function* `runApp()`"

- *R-dev: Ah, you got to do* `library(shiny)` *at the beginning of your script. And* `library(purrr)`, *and* `library(jsonlite)`, *and...*

For example here, the `library(purrr)` and `library(jsonlite)` will lead to
a NAMESPACE conflict on the `flatten()` function that can cause you some
debugging headaches (trust us, we have been there before). It would be cool
if we could have a {shiny} app that only imports specific functions from a
package, right?

We cannot stress enough that **dependencies matter**. You need to handle
them, and handle them correctly if you want to ensure a smooth deployment to production. This dependency management is native when you use a
package structure: the packages your application depends on are listed in the
DESCRIPTION, and the functions/packages you need to import are listed in the
NAMESPACE file.

C. It's split into functions

Third, let's say you are building a big application. Something with thousands of lines of code. You cannot build this large application by writing one or two files, as it is simply impossible to maintain in the long run or use on a daily basis. If we are developing a large application, we should split everything into smaller files. And maybe we can store those files in a specific directory.

This is what is done with a package, with the R/ folder.

D. It has documentation

Last but not least, we want our app to live long and prosper, which means we need to document it.

Documenting your {shiny} app involves explaining features to the end users and also to the future developers (chances are this future developer will be you). The good news is that using the R package structure helps you leverage the common tools for documentation in R:

- A README file that you will put at the root of your package, which will document how to install the package, and some information about how to use the package. Note that in many cases developers go for a .md file (short for markdown) because this format is automatically rendered on services like GitHub, GitLab, or any other main version control system.

- Vignettes are longer-form documentation that explains in more depth how to use your app. They are also useful if you need to detail the core functions of the application using a static document, notably for prototyping and/or for exchanging with the client. We will get back to Vignettes in Chapter 9 when we will talk about prototyping.

- Function documentation. Every function in your package should come with its own documentation, even if it is just for your future self. "Exported" functions, the one which are available once you run library(myapp), should be fully documented and will be listed in the package help page. Internal functions need less documentation, but documenting them is the best way to be sure you can come back to the app in a few months and still know why things are the way they are, what the pieces of the apps are used for, and how to use these functions.[1]

---

[1] {roxygen2} comes with a @noRd tag that prevents the documentation from being built. This allows you to still write the documentation using the same tags as the exported function, without the internal functions being documented in the end package. For example, that is why, by default, the modules built with {golem} version > 0.2.0 come with @noRd: you should document them, but chances are your do not need to export them.

- If needed, you can build a {pkgdown} website, that can either be deployed on the web or kept internally. It can contain installation steps for I.T., internal features use for developers, a user guide, etc.

E. It's tested

The other thing we need for our application to be successful in the long run is a testing infrastructure, so that we are sure we are not introducing any regression during development.

Nothing should go to production without being tested. Nothing.

Testing production apps is a broad question that we will come back to in another chapter, but let's talk briefly about why using a package structure helps with testing.

Frameworks for package testing are robust and widely documented in the R world, and if you choose to embrace the "{shiny} app as a package" structure, you do not have to put in any extra-effort for testing your application back-end: use a canonical testing framework like {testthat} (Wickham, 2021). Learning how to use it is not the subject of this chapter, so feel free to refer to the documentation, and see also Chapter 5 of the workshop: "Building a Package that Lasts"[2].

We will come back to testing in Chapter 11: "Build Yourself a Safety Net".

F. There is a native way to build and deploy it

Oh, and it would also be nice if people could get a `tar.gz` and install it on their computer and have access to a local copy of the app! Or if we could install that on the server without any headache!

When adopting the package structure, you can use classical tools to locally install your {shiny} application, i.e. as any other R package, built as a `tar.gz`. On a server, be it RStudio products or as Docker containers, the package infrastructure will also allow you to leverage the native R tools to build, install and launch R code.

More about this in Chapter 13.

{shiny} app as a package, a checklist

Let's sum up what we need for our application:

---

[2]https://speakerdeck.com/colinfay/building-a-package-that-lasts-erum-2018-workshop?slide=107

☐ **Metadata** and **dependencies**, which is what you get from the DESCRIPTION + NAMESPACE files of a package. Even more useful is the fact that you can do "selective namespace extraction" inside a package, *i.e.* you can say "I want this function from this package".

☐ A **bite-size code-base, aka functions**. This is done through .R files, stored in the R/ directory, which is the way a package is organized.

☐ **Documentation** can be done using Vignettes, Readme, and native R package documentation.

☐ **Testing toolkit**, done using the native R CMD Check and other packages like {testthat} (Wickham, 2021).

☐ **Installation process**, which is possible using the package infrastructure.

3.1.2   Resources

In the rest of this book, we will assume you are comfortable with building an R package. If you need to read some resources before continuing, feel free to have a look at these links:

- R packages[3]

- Building a Package that Lasts[4]

- Writing R Extensions[5]

- R Package Primer - a Minimal Tutorial[6]

3.2   Using {shiny} modules

Modules are one of the most powerful tools for building {shiny} applications in a maintainable and sustainable manner.

3.2.1   Why {shiny} modules?

Small is beautiful. Being able to properly cut a codebase into small modules will help developers build a mental model of the application (Remember "What is a complex {shiny} application?"). But what are {shiny} modules?

{shiny} modules address the namespacing problem in {shiny} UI (User Interface) and server logic, adding a level of abstraction beyond functions.

Modularizing Shiny app code (https://shiny.rstudio.com/articles/modules.html)

Let us first untangle this quote with an example about the {shiny} namespace problem.

A. The one million "Validate" buttons problem

A big {shiny} application usually requires reusing pieces of the UI/server, which makes it hard to name and identify similar inputs and outputs.

{shiny} requires its outputs and inputs to have a **unique id**. And, unfortunately, we cannot bypass that: when you send a plot **from R to the browser**, i.e. from the `server` to the `ui`, the browser needs to know exactly where to put this element. This "exactly where" is handled through the use of an `id`. Ids are not {shiny} specific: they are at the very root of the way web pages work. Understanding all of this is not the purpose of this chapter: just remember that {shiny} input and output ids **have** to be unique, just as any id on a web page, so that the browser knows where to put what it receives from R, and R knows what to listen to from the browser. The need to be unique is made a little bit complex by the way {shiny} handles the names, as it shares a global pool for all the id names, with no native way to use namespaces. Wait, namespaces?

Namespaces are a computer science concept created to handle a common issue: how to share the same name for a variable in various places of your program without them conflicting. In other words, how to use an object called `my_dataset` several times in the program, and still be sure that it is correctly used depending on the context.

R itself has a system for namespaces; this is what packages do and why you

can have `purrr::flatten` and `jsonlite::flatten` on the same computer
and inside the same script: the function names are the same, but the two exist
in different namespaces, and the behavior of both functions can be totally different as the symbol is evaluated inside two different namespaces. If you want
to learn more about namespaces, please refer to the 7.4 Special environments[7]
chapter from *Advanced R*, or turn to any computer science book: namespaces
are pretty common in any programming language.

That is what modules are made for: **creating small namespaces where
you can safely define ids without conflicting with other ids in the
app.** Why do we need to do that? Think about the number of times you
have created an "OK" or "validate" button. How have you been handling that
so far? By creating `validate1`, `validate2`, and so on and so forth. But if
you think about it, you are mimicking a namespacing process: a `validate` in
namespace 1, another in namespace 2.

Consider the following {shiny} application:

```
library(shiny)
ui <- function() {
  fluidPage(
    # Define a first sliderInput(),
    # with an id that we will postfix with `1`
    # in order to make it unique
    sliderInput(
      inputId = "choice1",
      label = "choice 1",
      min = 1, max = 10, value = 5
    ),
    # Define a first actionButton(),
    # with an id that we will postfix with `1`
    # in order to make it unique
    actionButton(
      inputId = "validate1",
      label = "Validate choice 1"
    ),
    # We define here a second sliderInput,
    # and need its id to be unique, so we
    # postfix it with 2
    sliderInput(
      inputId = "choice2",
      label = "choice 2",
      min = 1, max = 10, value = 5
```

---

[7]https://adv-r.hadley.nz/environments.html#special-environments

```
    ),
    # We define here a second actionButton,
    # and need its id to be unique, so we
    # postfix it with 2
    actionButton(
      inputId = "validate2",
      label = "Validate choice 2"
    )
  )
}

server <- function(input, output, session) {

  # Observing the first series of inputs
  # Whenever the user clicks on the first validate button,
  # the value of choice1 will be printed to the console
  observeEvent( input$validate1 , {
    print(input$choice1)
  })

  # Same as the first observeEvent, except that we are
  # observing the second series
  observeEvent( input$validate2 , {
    print(input$choice2)
  })
}

shinyApp(ui, server)
```

This, of course, is an approach that works. Well, it works as long as your codebase is small. But how can you be sure that you are not creating `validate6` on line 55 and another on line 837? How can you be sure that you are deleting the correct combination of UI/server components if they are named that way? Also, how do you work smoothly in a context where you have to scroll from `sliderInput("choice1")` to `observeEvent( input$choice1 , {})` which might be separated by thousands of lines?

B. Working with a bite-sized codebase

Build your application through multiple smaller applications that are easier to understand, develop and maintain, using {shiny} (Chang et al., 2021a) modules.

We assume that you know the saying that "if you copy and paste something more than twice, you should make a function". In a {shiny} application, how can we refactor a partially repetitive piece of code so that it is reusable?

Yes, you guessed right: using shiny modules. **{shiny} modules aim at three things: simplify "id" namespacing, split the codebase into a series of functions, and allow UI/Server parts of your app to be reused. Most of the time, modules are used to do the two first. In our case, we could say that 90% of the modules we write are never reused;[8] they are here to allow us to split the codebase into smaller, more manageable pieces.**

With {shiny} modules, you will be writing a combination of UI and server functions. Think of them as small, standalone {shiny} apps, which handle a fraction of your global application. If you develop R packages, chances are you have split your functions into series of smaller functions. With {shiny} modules, you are doing the exact same thing: with just a little bit of tweaking, you can split your application into a series of smaller applications.

3.2.2 When to use {shiny} modules

No matter how big your application is, it is always safe to start modularizing from the very beginning. The sooner you use modules, the easier downstream development will be. It is even easier if you are working with {golem}, which promotes the use of modules from the very beginning of your application.

"Yes, but I just want to write a small app, nothing fancy."

A production app almost always started as a small proof of concept. Then, the small PoC becomes an interesting idea. Then, this idea becomes a strategic asset. And before you know it, your not-that-fancy app needs to become larger and larger. So, you will be better off starting on solid foundations from the very beginning.

3.2.3 A practical walkthrough

An example is worth a thousand words, so let's explore the code of a very small {shiny} application that is split into modules.

A. Your first {shiny} module

Let's try to transform the above example (the one with two sliders and two action buttons) into an application with a module. Note that the following code will work only for {shiny} version 1.5.0 and after.

---

[8]Most of the time, pieces / panels of the app are too unique to be reused elsewhere.

```r
# Re-usable module
choice_ui <- function(id) {
  # This ns <- NS structure creates a
  # "namespacing" function, that will
  # prefix all ids with a string
  ns <- NS(id)
  tagList(
    sliderInput(
      # This looks the same as your usual piece of code,
      # except that the id is wrapped into
      # the ns() function we defined before
      inputId = ns("choice"),
      label = "Choice",
      min = 1, max = 10, value = 5
    ),
    actionButton(
      # We need to ns() all ids
      inputId = ns("validate"),
      label = "Validate Choice"
    )
  )
}

choice_server <- function(id) {
  # Calling the moduleServer function
  moduleServer(
    # Setting the id
    id,
    # Defining the module core mechanism
    function(input, output, session) {
      # This part is the same as the code you would put
      # inside a standard server
      observeEvent( input$validate , {
        print(input$choice)
      })
    }
  )
}


# Main application
library(shiny)
app_ui <- function() {
  fluidPage(
```

```
    # Call the UI  function, this is the only place
    # your ids will need to be unique
    choice_ui(id = "choice_ui1"),
    choice_ui(id = "choice_ui2")
  )
}


app_server <- function(input, output, session) {
    # We are now calling the module server functions
    # on a given id that matches the one from the UI
    choice_server(id = "choice_ui1")
    choice_server(id = "choice_ui2")
}


shinyApp(app_ui, app_server)
```

Let's stop for a minute and decompose what we have here.

The **server** function of the module (`mod_server()`) is pretty much the same as before: you use the same code as the one you would use in any server part of a {shiny} application.

The **ui** function of the module (`mod_ui()`) requires specific things. There are two new things: `ns <- NS(id)` and `ns(inputId)`. That is where the namespacing happens. Remember the previous version where we identified our two "validate" buttons with slightly different namespaces: `validate1` and `validate2`? Here, we create namespaces with the `ns()` function, built with `ns <- NS(id)`. This line, `ns <- NS(id)`, is added on top of all module UI functions and will allow building namespaces with the module id.

To understand what it does, let us try and run it outside {shiny}:

```
# Defining the id
id <- "mod_ui_1"
# Creating the internal "namespacing" function
ns <- NS(id)
# "namespace" the id
ns("choice")
```

```
[1] "mod_ui_1-choice"
```

And here it is, our namespaced `id`!

Each call to a module with `choice_server()` requires a different `id` argument that will allow creating various internal namespaces to prevent from id conflicts.[9] Then you can have as many `validate` ids as you want in your whole app, as long as this input has a unique id inside your module.

Note for {shiny} < 1.5.0

Released on 2020-06-23, the version 1.5.0 of {shiny} introduced a new way to write shiny modules, using a new function called `moduleServer()`. This new function was introduced to make the couple `ui_function` / `server_function` more obvious, where the old version used to require a `callModule(server_function, id)` call. This `callModule` notation is still valid (at least at the time of writing these lines), but we chose to go for the `moduleServer()` notation in this book. Most of the applications that are used as examples in this book have been built before this new function though, so when you will read their code, you will find the `callModule` implementation.

B. Passing arguments to your modules

{shiny} modules will potentially be reused and may need a specific user interface and inputs. This requires using extra arguments to generate the UI and server. As the UI and server are functions, you can set parameters that will be used to configure the internals of the result.

As you can see, the **app_ui** contains a series of calls to the `mod_ui(unique_id, ...)` function, allowing additional arguments like any other function:

```
mod_ui <- function(id, button_label) {
  ns <- NS(id)
  tagList(
    actionButton(ns("validate"), button_label)
  )
}
```

```
# Printing the HTML for this piece of UI
mod_ui("mod_ui_1", button_label = "Validate ")
mod_ui("mod_ui_2", button_label = "Validate, again")
```

```
<!-- The ids are "namespaced" with mod_ui_*-->
<button id="mod_ui_1-validate" type="button"
```

---

[9] Well, of course you can still have inner module id conflicts, but they are easier to avoid, detect, and fix.

```
class="btn btn-default action-button">Validate</button>

<!-- The ids are "namespaced" with mod_ui_*-->
<button id="mod_ui_2-validate" type="button"
class="btn btn-default action-button">Validate, again</button>
```

The **app_server** side contains a series of `mod_server(unique_id, ...)`, also allowing additional parameters, just like any other function.

As a live example, we can have a look at mod_dataviz.R[10] from the {tidytuesday201942} (Fay, 2021k) {shiny} application, available at `https://connect.thinkr.fr/tidytuesday201942/`. Figure 3.1 is a screenshot of this application.

This application contains 6 tabs, 4 of them being pretty much alike: a side bar with inputs, a main panel with a button, and the plot. This plot can be, depending on the tab, a scatterplot, a histogram, a boxplot, or a barplot.

This is a typical case where you should reuse modules: if several parts are relatively similar, it is easier to bundle it inside a reusable module, and condition the UI/server with function arguments.

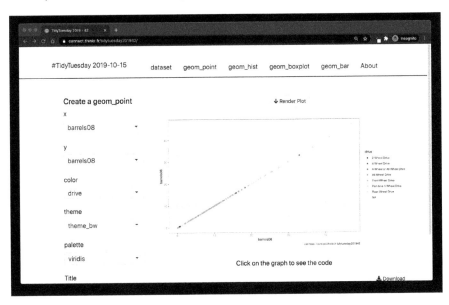

FIGURE 3.1: Snapshot of the {tidytuesday201942} {shiny} application.

Let's extract some pieces of this application to show how (and why) you would parametrize your module.

---

[10]`https://github.com/ColinFay/tidytuesday201942/blob/master/R/mod_dataviz.R#L17`

```r
mod_dataviz_ui <- function(
  id,
  type = c("point", "hist", "boxplot", "bar")
) {
  # Setting a header with the specified type of graph
  h4(
    sprintf( "Create a geom_%s", type )
  ),
  # [ ... ]
  # We want to allow a coord_flip only with barplots
  if (type == "bar"){
    checkboxInput(
      ns("coord_flip"),
      "coord_flip"
    )
  },
  # [ ... ]
  # We want to display the bins input only
  # when the type is histogram
  if (type == "hist") {
    numericInput(
      ns("bins"),
      "bins",
      30,
      1,
      150,
      1
    )
  },
  # [ ... ]
  # The title input will be added to the graph
  # for every type of graph
  textInput(
    ns("title"),
    "Title",
    value = ""
  )
}
```

And in the module server:

```r
mod_dataviz_server <- function(
```

```r
  input,
  output,
  session,
  type
) {
  # [ ... ]
  if (type == "point") {
    # Defining the server logic when the type is point
    # When the type is point, we have access to input$x,
    # input$y, input$color, and input$palette,
    # so we reuse them here
    ggplot(
      big_epa_cars,
      aes(
        .data[[input$x]],
        .data[[input$y]],
        color = .data[[input$color]])
    )  +
      geom_point()+
      scale_color_manual(
        values = color_values(
          1:length(
            unique(
              pull(
                big_epa_cars,
                .data[[input$color]])
            )
          ),
          palette = input$palette
        )
      )
  }
  # [ ... ]
  if (type == "hist") {
    # Defining the server logic when the type is hist
    # When the type is point, we have access to input$x,
    # input$color, input$bins, and input$palette
    # so we reuse them here
    ggplot(
      big_epa_cars,
      aes(
        .data[[input$x]],
        fill = .data[[input$color]]
      )
```

```
      ) +
      geom_histogram(bins = input$bins)+
      scale_fill_manual(
        values = color_values(
          1:length(
            unique(
              pull(
                big_epa_cars,
                .data[[input$color]]
              )
            )
          ),
          palette = input$palette
        )
      )
  }
}
```

Then, the UI of the entire application is:

```
app_ui <- function() {
  # [...]
  tagList(
    fluidRow(
      # Setting the first tab to be of type point
      id = "geom_point",
      mod_dataviz_ui(
        "dataviz_ui_1",
        type = "point"
      )
    ),
    fluidRow(
      # Setting the second tab to be of type point
      id = "geom_hist",
      mod_dataviz_ui(
        "dataviz_ui_2",
        type = "hist"
      )
    )
  )
}
```

And the app_server() of the application:

```r
app_server <- function(input, output, session) {
  # This app has been built before shiny 1.5.0,
  # so we use the callModule structure
  #
  # We here call the server module on their
  # corresponding id in the UI, and set the
  # parameter of the server function to
  # match the correct type of input
  callModule(mod_dataviz_server, "dataviz_ui_1", type = "point")
  callModule(mod_dataviz_server, "dataviz_ui_2", type = "hist")
}
```

3.2.4 Communication between modules

One of the hardest part of using modules is sharing data across them. There are at least three approaches:

- Returning a `reactive` function
- The "stratégie du petit r" (to be pronounced with a French accent of course)
- The "stratégie du grand R6"

A. Returning values from the module

One common approach is to return a `reactive` function from one module, **and pass it to another** in the general `app_server()` function.

Here is an example that illustrates this pattern.

```r
# Module 1, which will allow to select a number
choice_ui <- function(id) {
  ns <- NS(id)
  tagList(
    # Add a slider to select a number
    sliderInput(ns("choice"), "Choice", 1, 10, 5)
  )
}
```

```r
choice_server <- function(id) {
  moduleServer( id, function(input, output, session) {
```

```r
    # We return a reactive function from this server,
    # that can be passed along to other modules
    return(
      reactive({
        input$choice
      })
    )
  }
  )
}


# Module 2, which will display the number
printing_ui <- function(id) {
  ns <- NS(id)
  tagList(
    # Insert the number modified in the first module
    verbatimTextOutput(ns("print"))
  )
}

printing_server <- function(id, number) {
  moduleServer(id, function(input, output, session) {
    # We evaluate the reactive function
    # returned from the other module
    output$print <- renderPrint({
      number()
    })
  }
  )
}


# Application
library(shiny)
app_ui <- function() {
  fluidPage(
    choice_ui("choice_ui_1"),
    printing_ui("printing_ui_2")
  )
}

app_server <- function(input, output, session) {
  # choice_server() returns a value that is then passed to
  # printing_server()
  number_from_first_mod <- choice_server("choice_ui_1")
```

```
  printing_server(
    "printing_ui_2",
    number = number_from_first_mod
  )
}

shinyApp(app_ui, app_server)
```

This strategy works well, but for large {shiny} apps it might be hard to handle large lists of reactive outputs / inputs and to keep track of how things are organized. It might also create some reactivity issues, as a lot of `reactive` function calls is harder to control, or lead to too much computation from the server.

B. The "stratégie du petit r"

In this strategy, we **create a global reactiveValues list that is passed along through other modules**. The idea is that it allows you to be less preoccupied about what your module takes as input and what it outputs. You can think of this approach as creating a small, internal database that is passed along through all the modules of your application.

Below, we create a "global" (in the sense that it is initiated at the top of the module hierarchy) `reactiveValues()` object in the `app_server()` function. It will then go through all modules, passed as a function argument.

```
# Module 1, which will allow to select a number
choice_ui <- function(id) {
  ns <- NS(id)
  tagList(
    # Add a slider to select a number
    sliderInput(ns("choice"), "Choice", 1, 10, 5)
  )
}

choice_server <- function(id, r) {
  moduleServer(
    id,
    function(input, output, session) {
      # Whenever the choice changes, the value inside r is set
      observeEvent( input$choice , {
        r$number_from_first_mod <- input$choice
```

```r
    })

  }
 )
}

# Module 2, which will display the number
printing_ui <- function(id) {
  ns <- NS(id)
  tagList(
    # Insert the number modified in the first module
    verbatimTextOutput(ns("print"))
  )
}

printing_server <- function(id, r) {
  moduleServer(
    id,
    function(input, output, session) {
      # We evaluate the reactiveValue element modified in the
      # first module
      output$print <- renderPrint({
        r$number_from_first_mod
      })
    }
  )
}

# Application
library(shiny)
app_ui <- function() {
  fluidPage(
    choice_ui("choice_ui_1"),
    printing_ui("printing_ui_2")
  )
}

app_server <- function(input, output, session) {
  # both servers take a reactiveValue,
  #  which is set in the first module
  # and printed in the second one.
  # The server functions don't return any value per se
  r <- reactiveValues()
  choice_server("choice_ui_1", r = r)
```

```
  printing_server("printing_ui_2", r = r)
}

shinyApp(app_ui, app_server)
```

The good thing about this method is that whenever you add something in one module, it is immediately available in all other modules where r is present. The downside is that it can make it harder to reason about the app, as the input/content of the r is not specified anywhere unless you explicitly document it: the parameter to your server function being "r" only, you need to be a little bit more zealous when it comes to documenting it.

Note that if you want to share your module, for example in a package, you should document the structure of the r. For example:

```
#' @param r a `reactiveValues()` list with a
    `number_from_first_mod` element in it.
#' This `r$number_from_first_mod` will be
    printed to the `print` output.
```

C. The "stratégie du grand R6"

Similar to the "stratégie du petit r", we can create an R6 object, which is passed along inside the modules.

R6 objects, created using the package of the same name, are "traditional" object-oriented programming implementations in R. An R6 object is a data structure that can hold in itself data and functions. Its particularity is that if **it's modified inside a function, this modified value is kept outside the function in which it's called, making it a powerful tool to manage data across the application**.

As this R6 object is not a reactive object and is not meant to be used as such, uncontrolled reactivity of the application is reduced, thus reduces the complexity of handling chain reactions across modules. Of course, you need to have another special tool in your app to trigger elements. All this will be explained in detail in Chapter 15 of this book, and you can find an example of this pattern inside the {hexmake}[11] (Fay, 2021f) application.

If you are eager to know more about what {R6} is and how it works, we suggest the chapter on this subject in *Advanced R* (Wickham, 2019), R6[12].

---

[11]https://github.com/ColinFay/hexmake/blob/master/R/R6.R
[12]https://adv-r.hadley.nz/r6.html

D. Other approaches: About {tidymodules}

{tidymodules} (Larbaoui, 2021) is a package that helps in building shiny modules using an object-oriented paradigm, based on {R6}. It allows you to automatically take care of namespaces, and makes sharing data across modules easier. With this approach, as modules are objects, you can also define inheritance between modules.

The "Getting Started"[13] page for {tidymodules} offers a presentation of how you can build modules using this approach, so feel free to refer to it if you want to know more, and to dive into the article for the advanced features offered by this package.

3.3 Structuring your app

3.3.1 Business logic and application logic

A shiny application has two main components: the application logic and the business logic.

- **Application logic is what makes your {shiny} app interactive**: structure, buttons, tables, interactivity, etc. These components are not specific to your core business. You could use them for any other line of work or professional context. This has no other use case than your interactive application. It is not meant to be used outside your app, and you would not use them in a markdown report for instance.

- **Business logic includes the components with the core algorithms and functions that make your application specific to your area of work**. You can recognize these elements as the ones that can be run outside any interactive context. This is the case for specific computations and algorithms, custom plots or geom for {ggplot2} (Wickham et al., 2021a), specific calls to a database, etc.

These two components do not have to live together. And in reality, they should not live together if you want to keep your sanity when you are building an app. If you keep all components together in the same file, you will end up having to rerun the app from scratch and spend five minutes clicking everywhere just to be sure you have correctly set the color palette for the graph on the last tabPanel().

---

[13]https://opensource.nibr.com/tidymodules/articles/tidymodules.html

Trust us, we have been there, and it is not pretty.

What is the way to go? Extract the business function from the reactive functions. Literally. Compare this pattern:

```r
library(shiny)
library(dplyr)
# A simple app that returns a table
ui <- function() {
  tagList(
    tableOutput("tbl"),
    sliderInput("n", "Number of rows", 1, 50, 25)
  )
}

server <- function(input, output, session) {
  output$tbl <- renderTable({
    # Writing all the business logic for the table manipulation
    # inside the server
    mtcars %>%
      # [...] %>%
      # [...] %>%
      # [...] %>%
      # [...] %>%
      # [...] %>%
      top_n(input$n)
  })
}

shinyApp(ui, server)
```

To this one:

```r
library(shiny)
library(dplyr)

# Writing all the business logic for the table manipulation
# inside an external function
top_this <- function(tbl, n) {
  tbl %>%
    # [...] %>%
    # [...] %>%
```

```
  # [...] %>%
  # [...] %>%
    top_n(n)
}

# A simple app that returns a table
ui <- function() {
  tagList(
    tableOutput("tbl"),
    sliderInput("n", "Number of rows", 1, 50, 25)
  )
}


server <- function(input, output, session) {
  output$tbl <- renderTable({
    # We call the previously declared function inside the server
    # The business logic is thus defined outside the application
    top_this(mtcars, input$n)
  })
}

shinyApp(ui, server)
```

Both scripts do the exact same thing. The difference is that the second code can be easily explored without having to relaunch the app. You will be able to build a reproducible example to explore, illustrate, and improve `top_this()`. This function can be tested, documented, and reused outside the application. Moreover, this approach lowers the cognitive load when debugging: you either debug an application issue, or a business logic issue. You never debug both at the same time.

Even more, think about the future: how likely are the colors or the UI subject to change, compared to how likely the core algorithms are to change? As said in *The Art of Unix Programming*, "*Fashions in the look and feel of GUI toolkits may come and go, but raster operations and compositing are forever*" (Raymond, 2003). In other words, the core back-end, once consolidated, will potentially stay unchanged forever. On the other hand, the front-end might change: new colors, new graphic designs, new interactions, new visualization libraries, etc. Whenever this happens, you will be happy you have separated the business logic from the application logic, as you will have to change less code.

How to do that? Add your application logic in a file (typically, a module), and the business logic in another R script (typically starting with `fct_` or `utils_`).

You can even write the business logic inside another package, making these functions really reusable outside your application.

3.3.2 Small is beautiful (bis repetita)

There are a lot of reasons for splitting your application into smaller pieces, including the fact that it is easier to maintain, easier to decipher, and it facilitates collaboration.

There is nothing harder to maintain than a {shiny} app only made of a unique 1000-line long app.R file. Well, there still is the 10000-line long app.R file, but you get the idea. **Long scripts are almost always synonymous with complexity when it comes to building software**. Of course, small and numerous scripts do not systematically prevent codebase complexity, but they do simplify collaboration and maintenance, and divide the application logic into smaller, easier-to-understand bits of code.

In practice, big files are complex to handle and make development harder. Here is what happens when you work on an application for production:

- You will work during a long period of time (either in one run or split across several months) on your codebase. Hence, you will have to get back to pieces of code you wrote a long time ago.

- You will possibly develop with other developers. Maintaining a codebase when several people work on the same directory is already a complex thing: from time to time you might work on the same file separately, a situation where you will have to be careful about what and how to merge things when changes are implemented. It is almost impossible to work together on the same file throughout the project without losing your mind: even more if this file is thousands of lines long.

- You will implement numerous features. Numerous features imply a lot of UI and server interactions. In an app.R file containing thousands of lines, it is very hard to match the UI element with its server counterpart. When the UI is on line 50 and the server on line 570, you will be scrolling a lot when working on these elements.

3.3.3 Conventions matter

In this section you will find a suggestion for a naming convention for your app files that will help you and your team be organized.

Splitting files is good. Splitting files using a defined convention is better. Why? Because using a common convention for your files helps the other developers (and potentially you) to know exactly what is contained in a specific file, making it easier to navigate through the codebase, be it for newcomers or for developers already familiar with the software.

As developed in *Refactoring at Scale* (Lemaire, 2020), lacking a defined file structure when it comes to the codebase leads to slower productivity in the long run, notably when new engineers join the team: engineers with a knowledge of the file structure have learned how to navigate through the codebase, but new comers will find it hard to understand how everything is organized. And of course, in the long run, even developers with a knowledge of the structure can get lost, even more if they haven't worked on the project for months.

Because it's easier to maintain the status quo, instead of proactively beginning to organize related files [...], engineers instead learn to navigate the increasingly sprawling code. New engineers introduced to the growing chaos raise a warning flag and encourage the team to begin splitting up the code, but these concerns fall to deaf ears [...]. Eventually, the codebase reaches a critical mass where the persistent lack of organization has dramatically slowed productivity across the engineering team. Only then does the team take the time to draft a plan for grooming the codebase, at which point the number of variables to consider is far greater than had they made a concerted effort to tackle the problem months (or even years) earlier.

Refactoring at Scale (Lemaire, 2020)

Using a convention allows everyone to know where to look when debugging, refactoring, or implementing new features. For example, if you follow {golem}'s convention (which is the one developed in this section), you will know immediately that a file starting with mod_ contains a module. If you take over a project, look in the R/ folder, and see files starting with these three letters, you will know immediately that these files contain modules.

Here is our proposition for a convention defining how to split your application into smaller pieces.

First of all, put everything into an R/ folder. If you build your app using the {golem} framework, this is already the case. We use the package convention to hold the functions of our application.

The naming convention in {golem} is the following:

- `app_*.R` (typically `app_ui.R` and `app_server.R`) contain the top-level functions defining your user interface and your server function.

- `fct_*` files contain the business logic, which are potentially large functions. They are the backbone of the application and may not be specific to a given module. They can be added using {golem} with the `add_fct("name")` function.

- `mod_*` files contain a unique module. Many {shiny} apps contain a series of tabs, or at least a tab-like pattern, so we suggest that you number them according to their step in the application. Tabs are almost always named in the user interface, so that you can use this tab name as the file name. For example, if you build a dashboard where the first tab is called "Import", you should name your file `mod_01_import.R`. You can create this file with a module skeleton using `golem::add_module("01_import")`.

- `utils_*` are files that contain utilities, which are small helper functions. For example, you might want to have a `not_na`, which is `not_na <- Negate(is.na)`, a `not_null`, or small tools that you will be using application-wide. Note that you can also create `utils` for a specific module.

- `*_ui_*`, for example `utils_ui.R`, relates to the user interface.

- `*_server_*` are files that contain anything related to the application's back-end. For example, `fct_connection_server.R` will contain functions that are related to the connection to a database, and are specifically used from the server side.

Note that when building a module file with {golem}, you can also create `fct_` and `utils_` files that will hold functions and utilities for this specific module. For example, `golem::add_module("01_import", fct = "readr", utils = "ui")` will create R/mod_01_import.R, R/mod_01_import_fct_readr.R and R/mod_01_import_utils_ui.R.

Of course, as with any convention, you might occasionally feel like deviating from the general pattern. Your app may not have that many functions, or maybe the functions can all fit into one `utils_` file. But whether you have one or thousands of files, it is always a good practice to stick to a formalized pattern as much as possible.

4

Introduction to {golem}

The {golem} (Fay et al., 2021) package is a framework for building production-grade {shiny} applications. Many of the patterns and methodologies described in this book are linked to {golem} and packages from the golemverse[1]. Of course, all the advice developed in this book will still be valid even if you do not plan to use {golem}.

We have quickly introduced {golem} in the last chapter, and we will come back to this package many times in the following chapters. Let's start with an introduction to this package. Note that the version used at the time of writing this book is 0.3.1.

4.1 What is {golem}?

Note: The current version of {golem} used when writing this book is 0.3.1, and some of the features presented in this book might not be available if you are using an older version, or be a little bit different if you have a newer version. Feel free to browse the package NEWS.

{golem} is a **toolkit for simplifying the creation, development and deployment of a {shiny} application**. It focuses on building applications that will be sent to production, but of course starting with {golem} from the very beginning is also possible, even recommended: it is easier to start with {golem} than to refactor your codebase to fit into the framework.

The stable release can be found on CRAN and installed with:

```
install.packages("golem")
```

The {golem} development version can be found on GitHub and installed with:

---

[1]http://golemverse.org/

```r
remotes::install_github("Thinkr-open/golem")
```

The version of the package used while writing this book is:

```r
packageVersion("golem")
```

```
[1] '0.3.1'
```

The motivation behind {golem} is that building a proof-of-concept application is easy, but **things change when the application becomes larger and more complex, and especially when you need to send that app to production**. Until recently there has not been any real framework for building and deploying production-grade {shiny} apps. This is where {golem} comes into play: **offering {shiny} developers a toolkit for making a stable, easy-to-maintain, and robust production web application with R**. {golem} has been developed to abstract away the most common engineering tasks (for example, module creation, addition and linking of an external CSS or JavaScript file, etc.), so you can focus on what matters: building the application. Once your application is ready to be deployed, {golem} guides you through testing and brings tools for deploying to common platforms.

Some things to keep in mind before using {golem}:

- A {golem} application is contained inside a package. Knowing how to build a package is heavily recommended. The good news is also that everything you know about package development can be applied to {golem}.

- A {golem} app works better if you are working with `shiny modules`. Knowing how modules work is also recommended but not necessary.

4.2   Understanding {golem} app structure

A {golem} application is an R package. Having an R package architecture is perfectly suited for production-ready programs, as we developed in the previous chapter.

Let's focus on the architecture of the default {golem} app, and present the role that each file plays and how you can use (or not use) each of them.

You can create a {golem} project, here called golex, with RStudio "New project" creation or with command line.

```
golem::create_golem("golex")
```

The project will start with this specific architecture:

```
# Listing the files from the `golex` project using {fs}
fs::dir_tree("golex")
```

```
golex
+-- DESCRIPTION
+-- NAMESPACE
+-- R
|   +-- app_config.R
|   +-- app_server.R
|   +-- app_ui.R
|   \-- run_app.R
+-- dev
|   +-- 01_start.R
|   +-- 02_dev.R
|   +-- 03_deploy.R
|   \-- run_dev.R
+-- golex.Rproj
+-- inst
|   +-- app
|   |   \-- www
|   |       \-- favicon.ico
|   \-- golem-config.yml
\-- man
    \-- run_app.Rd
```

If you are familiar with building R packages, this structure will look familiar to you. And for a good reason: a {golem} app IS a package.

4.2.1 DESCRIPTION and NAMESPACE

The DESCRIPTION and NAMESPACE are **standard package files** (*i.e.* they are not {golem}-specific). In DESCRIPTION, you will add a series of metadata about your package, for example, who wrote the package, what is the package version, what is its goal, who to complain to if things go wrong, and also information about external dependencies, the license, the encoding, and so forth.

This `DESCRIPTION` file will be filled automatically by the first function you will run in `dev/01_start.R`, and by other functions from the `dev/` scripts. In other words, most of the time you will not interact with it directly, but through wrappers from {golem} and {usethis} (Wickham and Bryan, 2021) which are listed in the `dev` scripts.

The `NAMESPACE` file is the file you will NEVER edit by hand! **It defines how to interact with the rest of the package**: what functions to import and from which package and what functions to export, *i.e.* what functions are available to the user when you do `library(golex)`. This file will be built when running the documenting process in your R package, i.e. when doing `devtools::document()`, or more specifically in our case `golem::document_and_reload()`. This process will build the `man/` files and fill the `NAMESPACE`, by scanning the {roxygen} tags in your `.R` files.

If you want to learn more about these files, here are some resources you can refer to:

- Writing R Extensions—The DESCRIPTION file[2]
- Writing R Extensions—Package namespaces[3]
- R Packages—Package metadata[4]
- R Packages—Namespace[5]
- Building a package that lasts—eRum 2018 workshop[6]

4.2.2   R/

The `R/` folder is **the standard folder where you will store all your app functions**. When you start your project with {golem}, this folder is pre-populated with three .R files: `app_config.R`, `app_server.R`, `app_ui.R` and `run_app.R`.

During the process of building your application, all the core functionalities of your app will be stored in this `R/` directory, which is the standard way to store functions when using the R package framework. Note that these files are the

---

[2]https://cran.r-project.org/doc/manuals/r-release/R-exts.html#The-DESCRIPTION-file
[3]https://cran.r-project.org/doc/manuals/r-release/R-exts.html#Package-namespaces
[4]https://r-pkgs.org/description.html#
[5]https://r-pkgs.org/namespace.html
[6]https://speakerdeck.com/colinfay/building-a-package-that-lasts-erum-2018-workshop

"core" features of your application itself, and that other .R files also exists. For example, when you will need to deploy your application on RStudio platforms, {golem} will create an app.R at the root of your directory.[7] The dev/ folder also contains .R scripts, and they are inside this folder as they should not live inside the R/ folder: they are utilitarian files used during development, not core functionalities of your application.

Inside these .R files, contained inside the R/ folder, you will find the content of your modules (the one added with golem::add_modules()) and the utilitarian/business logic functions, built with golem::add_utils() and golem::add_fct().

Note also that this folder must not contain any sub-folders.

app_config.R

```r
#' Access files in the current app
#'
#' NOTE: If you manually change your package
#' name in the DESCRIPTION, don't forget to change it here too,
#' and in the config file. For a safer name change mechanism,
#' use the `golem::set_golem_name()` function.
#'
#' @param ... character vectors, specifying subdirectory
#' and file(s) within your package.
#' The default, none, returns the root of the app.
#'
#' @noRd
app_sys <- function(...){
  system.file(..., package = "golex")
}
```

```r
#' Read App Config
#'
#' @param value Value to retrieve from the config file.
#' @param config GOLEM_CONFIG_ACTIVE value.
#' If unset, R_CONFIG_ACTIVE.  If unset, "default".
#' @param use_parent Logical,
#' scan the parent directory for config file.
#'
```

---

[7]{golem} will automatically add this file to the .Rbuildignore file, i.e. make it be ignored by the package build process.

```
#' @noRd
get_golem_config <- function(
  value,
  config = Sys.getenv(
    "GOLEM_CONFIG_ACTIVE",
    Sys.getenv(
      "R_CONFIG_ACTIVE",
      "default"
    )
  ),
  use_parent = TRUE
){
  config::get(
    value = value,
    config = config,
    # Modify this if your config file is somewhere else:
    file = app_sys("golem-config.yml"),
    use_parent = use_parent
  )
}
```

The app_config.R file contains internal mechanics for {golem}, notably for referring to values in the inst/ folder, and to get values from the config file in the inst/ folder. Keep in mind that if ever you need to change the name of your application, you will need to change it inside the DESCRIPTION, but also inside the app_sys() function. To make this process easier, you can use the golem::set_golem_name(), which will perform both these actions, plus setting the name inside the config file.

app_server.R

```
#' The application server-side
#'
#' @param input,output,session Internal parameters for {shiny}.
#'     DO NOT REMOVE.
#' @import shiny
#' @noRd
app_server <- function( input, output, session ) {
  # Your application server logic

}
```

The `app_server.R` file **contains the function for the server logic**. If you are familiar with the classic "ui.R/server.R" approach, this function can be seen as a replacement for the content of the function you have in your `server.R`.

Building a complex {shiny} application commonly implies using {shiny} modules. If so, you will be adding there a series of `callModule()`, the ones you will get on the very bottom of the file created with `golem::add_module()`.

You will also find global elements from your server-logic: top-level `reactiveValues()`, connections to databases, setting options, and so forth.

app_ui.R

```r
#' The application User-Interface
#'
#' @param request Internal parameter for `{shiny}`.
#'     DO NOT REMOVE.
#' @import shiny
#' @noRd
app_ui <- function(request) {
  tagList(
    # Leave this function for adding external resources
    golem_add_external_resources(),
    # Your application UI logic
    fluidPage(
      h1("golex")
    )
  )
}
```

This piece of the `app_ui.R` is designed to **receive the counterpart of what you put in your server**. Everything here is to be put after the # Your application UI logic line. Just as with their server counterparts, the UI side of these elements are the ones from the bottom of the file you are creating with `golem::add_module()`.

By default, {golem} uses a `fluidPage()`, which is the most commonly used {shiny} (Chang et al., 2021a) template. If ever you want to use `navBarPage()`, this is where you will define it: replace one with the other, and you will be good to go. You can also define any other template page, for example with an `htmlTemplate()`. For an example of an application built using an `htmlTemplate`, please visit engineering-shiny.org/grayscale/[8], or engineering-

---

[8]https://engineering-shiny.org/grayscale/

shiny.org/golemhtmltemplate/[9]: both these applications are built on top of an external html template.

If you're tempted to do that, be aware that `fluidPage()` comes with a series of CSS/JS elements, and if you plan on not using a default {shiny} `*Page()` function, you will need to add your own CSS.

```
#' Add external Resources to the Application
#'
#' This function is internally used to add external
#' resources inside the Shiny application.
#'
#' @import shiny
#' @importFrom golem add_resource_path activate_js
#' @importFrom golem favicon bundle_resources
#' @noRd
golem_add_external_resources <- function(){

  add_resource_path(
    'www', app_sys('app/www')
  )

  tags$head(
    favicon(),
    bundle_resources(
      path = app_sys('app/www'),
      app_title = 'cloop'
    )
    # Add here other external resources
    # for example, you can add
    # shinyalert::useShinyalert()
  )
}
```

The second part of this file contains the `golem_add_external_resources()` function, which is used to add, well, external resources. You may have noticed that this function is to be found above in the file, in the `app_ui()` function. This function is used for **linking to external files inside your applications**: notably the files you will create with `golem::add_css_file()` and friends.

In `golem_add_external_resources()`, you can also define a custom resourcesPath. The first line (the one with `add_resource_path()`) is the

---

[9]https://engineering-shiny.org/golemhtmltemplate/

one allowing the `inst/app/www` folder to be mounted and be available at `www` with your app when you launch it. That link makes it possible for {golem} to bundle the CSS and JavaScript files automatically.

The other part of this function, starting with `tags$head`, creates a `<head>` tag for your application. This `<head>` tag is a pretty standard tag, which is used in HTML to define a series of metadata about your app. **The last part of this function, the one with `bundle_resources()`, links all the CSS and JavaScript files contained in `inst/app/www` to your application, so you don't have to link them manually.**

And finally, if you want to add other elements to the `<head>` of your application (for example, by calling `shinyalert::useShinyalert()` or `cicerone::use_cicerone()` as in {hexmake}),[10] you can add these calls after the `bundle_resources()` function. Note that as all these elements are inside a `tags$head()`, they are to be treated as a list, so separated by commas.

run_app.R

```r
#' Run the Shiny Application
#'
#' @param ... arguments to pass to golem_opts.
#' See `?golem::get_golem_options` for more details.
#' @inheritParams shiny::shinyApp
#'
#' @export
#' @importFrom shiny shinyApp
#' @importFrom golem with_golem_options
run_app <- function(
  onStart = NULL,
  options = list(),
  enableBookmarking = NULL,
  uiPattern = "/",
  ...
) {
  with_golem_options(
    app = shinyApp(
      ui = app_ui,
      server = app_server,
      onStart = onStart,
      options = options,
```

---

[10]See https://github.com/ColinFay/hexmake/blob/master/R/app_ui.R#L35

```
      enableBookmarking = enableBookmarking,
      uiPattern = uiPattern
    ),
    golem_opts = list(...)
  )
}
```

The run_app() function is the one that you will use to launch the app.[11]

The body of this function is wrapped inside with_golem_options(), which allows you to pass arguments to the run_app() function, which can be called later on with golem::get_golem_options(). **The idea here is that you can pass arguments to this function, and that arguments will be later used inside your application to display a specific version of the application.** Using this with_golem_options() function simplifies the parameterization of {shiny} applications, be it during development, when deployed on a server, or when shared as a package.

Here are some examples of what you can pass to your shiny application using this pattern:

- run_app(user_country = "france") and run_app(user_country = "germany") to launch the application and show the data for a specific country.

- run_app(with_mongo = TRUE) to launch the application with or without a MongoDB back-end (example taken from {hexmake}).

- run_app(dataset = iris) will make the dataset available with golem::get_golem_options("dataset"), so your user can launch the function from their package using a dataset they have created/loaded

4.2.3 golem-config

app_config.R

Inside the R/ folder is the app_config.R file. This file is designed to handle two things:

- app_sys() is a wrapper around system.file(package = "golex"), and allows you to quickly refer to the files inside the inst/ folder. For example, app_sys("x.txt") points to the inst/x.txt file inside your package.

---

[11]Very technically speaking, it is the print() from the object outputed by run_app() that launches the app, but this is another story.

- `get_golem_config()` helps you manipulate the config file located at `inst/golem-config.yml`.

Manipulating `golem-config.yml`

Here is what the default config file looks like:

```
default:
  golem_name: golex
  golem_version: 0.0.0.9000
  app_prod: no
production:
  app_prod: yes
dev:
  golem_wd: !expr here::here()
```

It is based on the `{config}` (Allaire, 2020) format, and allows you to define contexts, with values associated with these specific contexts. For example, in the default example:

- `default.golem_name`, `default.golem_version`, and `default.app_prod` are usable across the whole life of your golem app: while developing, and also when in production.
- `production.app_prod` might be used for adding elements that are to be used once the app is in production.
- `dev.golem_wd` is in a dev config because **the only moment you might reliably use this config is while developing your app.** Use the `app_sys()` function if you want to rely on the package path once the app is deployed.

These options are globally set with:

```
# This functions sets all the default options for your project
set_golem_options()
```

```
  Setting {golem} options in `golem-config.yml`
  Setting `golem_wd` to /Users/colin/golex
You can change golem working directory with
set_golem_wd('path/to/wd')
  Setting `golem_name` to golex
  Setting `golem_version` to 0.0.0.9000
  Setting `app_prod` to FALSE
  Setting {usethis} project as `golem_wd`
```

The functions reading the options in this config file are:

```r
# Get the values from the config file
get_golem_name()
```

```
[1] "golex"
```

```r
get_golem_wd()
```

```
[1] "/Users/colin/golex"
```

```r
get_golem_version()
```

```
[1] "0.0.0.9000"
```

You can set these with:

```r
# Get the values in the config file
set_golem_name("this")
```

```
Setting `golem_name` to this
```

```r
set_golem_version("0.0.1")
```

```
Setting `golem_version` to 0.0.1
```

```r
# Get the values from the config file
get_golem_name()
```

```
[1] "this"
```

```
get_golem_version()
```

```
[1] "0.0.1"
```

If you are already familiar with the {config} package, you can use this file just as any config file.

{golem} comes with an amend_golem_config() function to add elements to it.

```
# Add a key in the default configuration
amend_golem_config(
  key = "MONGODBURL",
  value = "localhost"
)
# Add a key in the production configuration
amend_golem_config(
  key = "MONGODBURL",
  value = "0.0.0.0",
  config = "production"
)
```

In R/app_config.R, you will find a get_golem_config() function that allows you to retrieve config from this config file:

```
# Retrieve the value of `where`
get_golem_config(
  "MONGODBURL"
)
```

```
[1] "localhost"
```

```
get_golem_config(
  "MONGODBURL",
  config = "production"
)
```

```
[1] "0.0.0.0"
```

You can also use an environment variable (default {config} behavior):

```
Sys.setenv("GOLEM_CONFIG_ACTIVE" = "production")
get_golem_config(
  "MONGODBURL"
)
```

```
[1] "0.0.0.0"
```

The good news is that if you don't want/need to use {config}, you can safely ignore this file, just leave it where it is: it is used internally by the {golem} functions.

golem_config vs golem_options

There are two ways to configure golem apps:

- The golem_opts in the run_app() function
- The golem-config.yml file

The big difference between these two is that the golem options from run_app() are meant to be configured during runtime: you will be doing run_app(val = "this"), whereas the golem-config is meant to be used in the back-end, and will not be linked to the parameters passed to run_app() (even if this is technically possible, this is not the main objective).

It is also linked to the GOLEM_CONFIG_ACTIVE and R_CONFIG_ACTIVE environment variables.

The idea is also that the golem-config.yml file is shareable across {golem} projects (golem_opts are application specific), and will be tracked by version control systems.

For example, let's imagine we want to deploy the {hexmake} application on two RStudio Connect instances, but that both need a different MongoDB configuration when it comes to port, db, and collection name. To do that, you can take several approaches:

- Set these values as run_app() parameters, but that means that you have to maintain one app.R for each server to which you will deploy.
- Set everything as environment variables, but that means that you have to do it for every server, and that there is no centralized way to keep track of these variables.

- Set the values in **golem-config.yaml**, and then set a value for the **GOLEM_CONFIG_ACTIVE** environment variable in the environment in which the app is deployed.

This last solution is a convenient one if you want to easily re-deploy your application on various servers without having to (re)set the values for each environment. Note, though, that it shouldn't be used to store sensitive data (for example users and passwords).

Here is the config file that would illustrate what we just said (we have removed the other golem-related entries for the sake of clarity):

```
default:
  url: mongo
  mongoport: 12345
  mongodb: users
  mongocollecton: hex
server1:
  url: mongo
  mongoport: 6543
  mongodb: users
  mongocollecton: hex
server2:
  url: mongo
  mongoport: 9876
  mongodb: shiny
  mongocollecton: hexmake
server2:
  url: 127.0.0.1
  mongoport: 3214
  mongodb: connect
  mongocollecton: app1
```

Using this configuration file, you can then deploy the very same app on the two servers, and configure what is going to be read by the application by setting an environment variable inside the RStudio Connect interface, as shown in Figure 4.1.

4.2.4 inst/app/www/

The inst/app/www/ folder contains all files that are made available **at application run time**. Any web application has external files that allow it to

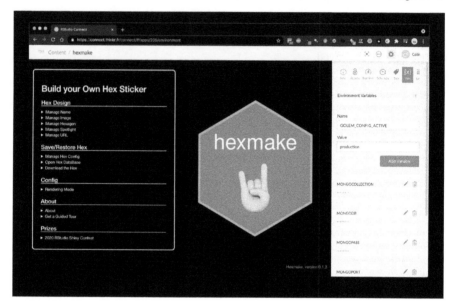

FIGURE 4.1: Setting an environment variable in RStudio Connect.

run.[12] For example, {shiny} and its fluidPage() function bundles a series
of CSS and JavaScript files, notably the Bootstrap library, or jQuery. These
external files enhance your app: CSS for the design part and JavaScript for the
interactive part (more or less). On top of that, you can add your own files: your
own design with CSS, or your own JavaScript content (as we will see in the
last chapters of this book). In order to work, you have to include a link to these
files somewhere in the UI. This is what golem_add_external_resources() is
made for: linking the external resources that you will build with the following
functions.

- golem::add_css_file()
- golem::add_js_file()
- golem::add_js_handler()
- golem::use_external_css_file()
- golem::use_external_js_file()
- golem::use_favicon()

Be aware that these files are available under www/ at **application run
time**, *i.e.* the www/ folder is available via your browser, not via R when it
runs/generates your application. In other words, you can use the www prefix
in the HTML generated in your UI, which is read by your browser, not from

---

[12]Some web pages do not need any external sources, as they do not have any design and
are plain HTML, but generally speaking we will not call this format a web application.

the R/server side. If you want to link to a file that is read during **application generation**, you will need to use the `app_sys()` function with, for example, `includeMarkdown( app_sys("app/www/howto.md") )`.

We encourage you to add any new external file (e.g. pictures) in the `inst/app/www` folder, so that you can later use it in the UI with the common `www` prefix. Another common pattern would be:

- Adding images in `inst/app/img`
- Calling `addResourcePath( 'img', system.file('app/img', package = 'golex') )`
- Adding elements to your UI with `tags$img(src = "img/name.png")`

4.2.5 dev/

The `dev/` folder is to be used as a **notebook for your development process: you will find here a series of functions that can be used throughout your project**.

The content of these files are specific to {golem} here, but the concept of using a script to store all development steps is not restricted to a {shiny} application: it could easily be done for any package, and this is something we recommend that you do. The functions inside these files are the ones used to do some setup, like `usethis::use_mit_license()` or `usethis::use_vignette("my-analysis")`, and add testing infrastructure, like `usethis::use_test("my-function")` or `devtools::check()`. You will also find functions to populate the application like `golem::add_module("my-module")` or `golem::add_js_file("my-script")`. And finally, there are functions you will need once your application is ready: `pkgdown::build_site()`, `rhub::check_for_cran()` or `golem::add_dockerfile()`.

We will come back to these files later in this book when we describe in more depth the {golem} workflow.

4.2.6 man/

The `man/` folder includes **the package documentation**. It is a common folder automatically filled when you document your app, notably when running the `dev/run_dev.R` script and the `document_and_reload()` function.

Building documentation for a package is a widely documented subject, and if you want to know more about documentation and how to build it, here are some external links:

- R Packages - Object documentation[13]
- Introduction to roxygen2[14]
- Building a package that lasts—eRum 2018 workshop[15]

---

[13]http://r-pkgs.had.co.nz/man.html
[14]https://cran.r-project.org/web/packages/roxygen2/vignettes/roxygen2.html
[15]https://speakerdeck.com/colinfay/building-a-package-that-lasts-erum-2018-workshop

5

The Workflow

Building a robust, production-ready web application will be made easier by following a given workflow. The one we are advocating is divided in five steps:

- Design
- Prototype
- Build
- Strengthen
- Deploy

In this chapter, we will give an brief overview of the different steps: the rest of the book will cover each of these steps in more depth.

Of course, as with any workflow, this one is not a one-size-fits-all solution: all projects are unique, with technical requirements, specific planning and team of coder(s). But we think that following this workflow will help you get good habits when it comes to structuring your application project, even more if you know from day one that the application you are going to work on is a large application, whether in terms of codebase, complexity, or time.

Note that the ideas behind this workflow, and its process, could be used outside of a {shiny} project: it can be applied to any coding project, even outside of the R world. Of course, the tools presented in this book are R and {shiny} specific, but the general ideas can be bootstrapped to be used outside of this context.

5.1 Step 1: Design

The first part of the workflow is the **design** part.

This very first step is the one that happens before starting to code: it is the one where you are thinking about the general implementation and features of the application, and where you build the general roadmap for the coding process. During the process of designing, you will define how the application will be built: somewhere between users' dreams, what is technically possible, and the time you have to build the application.

This first step is not {shiny} or R specific, it is something software engineers do for any software or web application: discuss with the clients,[1] the end users, and the developers who will work on the project. The idea with this first step is to get a clear idea of what everybody involved in the project wants/is able to do:

- From the client/end user's point of view, this step involves working on getting a clear idea of what they want the application to do, and to confront this view with the developers to evaluate what is possible to do, how much time it will take to implement desired features, etc.

- From the developer team point of view, this step also involves getting a clear idea of what the client is asking, in other words it involves translating the requirements to technical specifications. For example, the client might write something like "Save the plot inside a database so that we can search for them later on": from an application user point of view, this is a clear feature, from a developer point of view, this requirement can be translated in many ways.

This first step actually implies a lot of thinking before coding. The main goal of this step is to spend time thinking about the application while you still do not have anything implemented, so that you do not discover blocking elements once it is too late, or at least once you already have written a lot of code. We have all been in a situation during a project where we tell ourselves: "I wish I had known this sooner": working on designing the application before building it helps lowering the chances for this kind of bad surprise.

This first part of the workflow will span three chapters:

- "UX Matters", Chapter 6, is a chapter where we introduce the concepts of "User eXperience" (UX), and why it is a crucial concept when you are building your application. This chapter will cover the importance of simplicity when creating web applications, the danger of trying to implement too many components (aka "feature creep"), and finally we will introduce some general rules about web accessibility. These topics are vast topics, and a lot of literature and online resources exist for all these subjects: further readings and resources are linked inside each section.

- Chapter 7, "Don't rush into coding" underlines why "coding first" might not be the best strategy when it comes to building a production application. We will also quickly introduce concept maps, and list some of

---

[1] We use the term "client" in a loose sense, meaning the person(s) who is/are ordering the application.

the common questions you might want to ask the people involved in the project.

- Finally, this first part of the workflow covers a gentle introduction to CSS, which might be a crucial skill to master when it comes to sending an application to production: either your clients already have a CSS template that they want to include in the application, or they want their application to have the color and design that match the one from the company. Also, when building a professional application, chances are that you will want your app to stand out from the crowd: hence a little bit of CSS. This part is included in the design part because it is something that you might want to think about from the very beginning: for example, some companies have pre-existing {shiny} templates, they might want to include specific fonts, logo, icons, etc. These are things better known before starting to code: it is easier to start working inside a {shiny} template than migrating an existing code to a template.

5.2 Step 2: Prototype

The **prototype** part is the one during which you will build the front-end and the back-end, but separately.

As you may know, a {shiny} application is an interface (the front-end, or "UI") used to communicate information to the end users that are computed on the server side (the back-end, or "server").

To start on solid ground, you need to build the two (front and back) separately.

On one hand, work on the general appearance, without working on any actual algorithmic implementation: position of the inputs and outputs, general design, interactions, etc.; everything that does not rely on computation on the back-end. This "UI first" approach will be made possible for {shiny} with notably one package, {shinipsum} (Fay and Rochette, 2021b).

On the other hand, you (or someone from your team), will be working on building the back-end logic, which comprises the actual outputs that are going to be displayed, the algorithm that will compute results, and all the elements that do not need an interactive runtime to work. For this point, you can use what we call a "Rmd-first" approach, by combining R functions with the writing of vignettes that describe the internals of the application. This part of the workflow will be developed in two chapters:

- Chapter (8), "Setting up for success with {golem}", will cover the basics of getting started with the {golem} package so that you can start your prototyped application with solid foundation.

- Chapter (9), "Building an "ipsum-app"" will cover the importance of prototyping when it comes to building applications, then present {shinipsum} and {fakir}, and finally will introduce how you can use the "Rmd First" methodology to prototype your application back-end.

5.3 Step 3: Build

The **build** part is the one where you will combine the business (or back-end) logic with the front-end. In this third part, you will work on the core engine of the application, making the business logic work inside the interactive logic of your application.

This step of the workflow is cover in *Building app with {golem}* (10), a chapter that presents the various functions you can use to build your application, *i.e* the one you will be using to combine your back-end and front-end.

In this step, we will cover:

- How to handle dependencies in your project, *i.e* how to use external libraries inside your project
- How to organize modules and functions inside your project
- How to add tests for the back-end of your application (testing will be covered in more depth in Step 4)
- How to document your application and its codebase, and how to add code coverage and continuous integration
- How to leverage the internal `dev` functions from {golem} to modify the behavior of specific functions based on an `option()`

5.4 Step 4: Strengthen

The **strengthen** part covers how to ensure your application is immortal, in the sense that we defined in Chapter 1 of this book.

In this part, we will go through unit tests, reproducible development environments, version control, and continuous integration in the context of {shiny} applications. Building a solid testing suite is crucial to the success of a project, as it allows a project to be stable in the long run, be it when you will want to add new feature or refactor existing code:

Refactoring requires we be able to confidently ensure that behavior remains identical at every iteration. We can increase our confidence that nothing has changed by writing a suite of tests (unit, integration, end-to-end), and we should not seriously consider moving forward with any refactoring effort until we've established sufficient test coverage.

Refactoring at Scale (Lemaire, 2020)

This step of the workflow will span over chapters.

- The first one, "Build yourself a safety net" (Chapter 11), details how to build a testing environment for your {shiny} application, be it for testing the back-end or the front-end. In this chapter, you will be introduced to {testthat} for testing your application back-end, tools that are more linked to testing the front-end like NodeJS puppeteer module, {shinytest} and {crrry} for testing interactive logic, {shinyloadtest} and {dockerstats} for testing your application load. This chapter will also cover {renv} and Docker, two essential tools for developing in a reproducible environment.

- In Chapter 12, "Version Control", you will be introduced to git and to automated testing using continuous integration (CI) platforms like Travis CI or GitHub Actions.

5.5 Step 5: Deploy

To **deploy** is to send your application into production once it is built.

Being exhaustive here would be an impossible task: there are countless ways to make your application accessible to its targeted users, but we will try to cover some basics in this part. And of course, where and how you will be deploying your application depends on a lot of parameters. For example, who are the end users, and how do they want to use your application? If the end users are familiar with R and use it on a daily basis, they might be looking for an application that runs with `library(app)`, *i.e* they need the application to be available as an R package they can install on their machine. If the end users are not coders, they might need the application to be available only as a web application, so they just have to open a browser and navigate to a URL. Both these cases raise other questions: how can you make the package available on a repository so that R users can get it with `install.packages()`? If the application is to be made available on a URL, how will it be deployed? What deployment server is available to you, or to the company ordering the application? These questions (and more) will be covered in the **deploy** part of this book.

In this part, we will present a series of methods to prepare your application to be deployed on various environments, notably:

- Sharing your application as a package so that it can be installed manually, through GitHub, or shared on a package repository like the CRAN or BioConductor
- Sending it to an RStudio platform
- Building a Docker image to serve your app on a cloud provider

This step of the workflow is covered in the Chapter 13, "Deploy your application".

Part II

Step 1: Design

6

UX Matters

Let's state the truth: no matter how complex and innovative your back-end is, your application is bad if your user experience (UX) is bad. That's the hard truth. We have a natural tendency, as R-coders, to be focused on the back-end, i.e. the server part of the application, which is perfectly normal—chances are you did not come to R to design front-ends.[1]

However, **if people cannot understand how to use your application, or if your application front-end does not work at all, your application is not successful no matter how innovative and incredible the computation algorithms in the back-end are.**

As you are building a complex, production-grade {shiny} application, do not underestimate the necessity for a successful front-end - it is, after all, the first thing (and probably the only thing) that the end users of your web application will see. However, our natural back-end/server logic as R developers can play against us in the long run - **by neglecting the UI and the UX, you will make your application less likely to be adopted among your users, which is a good way to fail your application project.**

6.1 Simplicity is gold

Simplify, then add lightness.

Colin Chapman CBE, Founder of Lotus Cars (https://www.lotuscars.com/lotus-philosophy/)

---

[1] The front-end is the visual part of your application - the one your user interacts with - as opposed to the back-end, which is what is installed on the server, the part the end user does not see. In {shiny}, front-end corresponds to the UI, while back-end, to the server.

Aiming for simplicity is a hard thing, but some rules will help you build a better UX, paving the way for a successful application.

There are mainly two contexts where you will be building a web app with R: for professional use (*i.e.,* people will rely on the app to do their job), or for fun (*i.e.,* people will just use the app as a distraction).

But both cases have something in common: people will want the app to be usable, **easily** usable.

If people use your app in a professional context, they do not want to fight with your interface, read complex manuals, or lose time understanding what they are supposed to do and how they are supposed to use your application, at least when it comes to the core usage of the application. This core usage needs to be "self-explanatory", in the sense that, **if possible, the main usage of the application does not require reading the manual**; On the other hand, more advanced/rarely used features will need more detailed documentation.

In other words, they want an efficient tool, something that - beyond being accurate - is easy to grasp. In a professional context, when it comes to "business applications", remember that the quicker you understand the interface, the better the user experience. Think about all the professional applications and software that you have been ranting about during your professional life, all these cranky user interfaces you did not understand and/or need to re-learn every time you use them. You do not want your app to be one of these applications.

On the other hand, if users open your app for fun, they are not going to fight against your application; they are just going to give up if the app is too complex to use. Even a game has to appear easy to use when the users open it.

In this section, we will review two general principles: the "don't make me think" principle, which states that **interfaces should be as self-explanatory as possible**, and the "rule of least surprise", which states that elements should behave the way they are commonly expected to behave. These two rules aim at solving one issue: the bigger the cognitive load of your app, the harder it will be for the end user to use it on a daily basis.

6.1.1   How we read the web: Scanning content

One big lie we tell ourselves as developers is that the end user will use the app the way we designed it to be used (though to be honest, this is not true for any software). We love to think that when faced with our app, the users will carefully read the instructions and make a rational decision based on careful examination of the inputs before doing what we expect them to do. But the harsh truth is, that it is not what happens.

First of all, users rarely carefully read all the instructions: they **scan** and perform the first action that more or less matches what they need to do, i.e., they **satisfice** (a portmanteau of satisfy and suffice); a process shown in Figure 6.1. Navigating the web, users try to optimize their decision, not by making the decision that would be "optimal", but by doing the first action that is sufficiently satisfactory in relevance. They behave like that for a lot of reasons, but notably because they want to be as quick as possible on the web, and because the cost of being wrong is very low most of the time - even if you make the wrong decision on a website, chances are that you are just a "return" or "cancel" button away from canceling your last action.

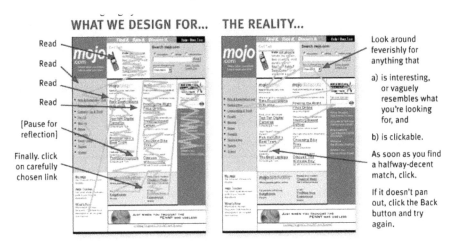

FIGURE 6.1: How we design a web page versus how a user will really scan it. From (Krug, 2014).

For example, let's have a look at the user interface of {hexmake} (Fay, 2021f), a {shiny} app for building hex stickers, available at https://connect.thinkr. fr/hexmake/ (see Figure 3 for a screenshot of this application).

What will be your reading pattern for this application? What is the first thing you will do when using this app?

There is an inherent logic in the application: each sub-menu is designed to handle one specific part of your sticker. The second menu is the one used to download the sticker, and the last menu is the one used to open the "how to" of the app. When opening this app, will your first move be to open the "How to"? Will you open all the sub-menus and select the most "logical" one to start with? Chances are that by reading this line, you think you will do that. But in reality, we behave less rationally than we'd like to think. What we do most of the time is click on the first thing that matches what we are here to do. For example, most of the time we will first change the package name or upload an image before even opening the "about" section of this app.

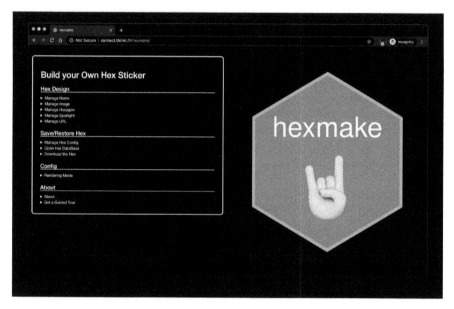

FIGURE 6.2: Snapshot of the {hexmake} {shiny} application on https:
//connect.thinkr.fr/hexmake/.

Once users have scanned the page, they perform the first action that seems reasonable, or as coined in "Rational Choice and the Structure of the Environment" by Herbert A. Simon, **"organisms adapt well enough to 'satisfice'; they do not, in general, optimize.".** In other words, **"As soon as we find a link that seems like it might lead to what we're looking for, there's a very good chance that we'll click it"** (*Don't Make Me Think*, (Krug, 2014)).

What that also means is that user might perform what you'd expect to be "irrational" choices. As they are scanning your application, they might do something unexpected, or use a part of your app in a way that you would not expect it to be used. For example, if you are creating an app that is designed to take input data that has to be filled in following a specific form, you **need** to check that this requirement is fulfilled, or you will end up debugging errors on uncommon entries.

This is a pretty common thing about apps and about software in general: you have to expect users to use your product in ways you would not have expected, in ways that might seem absurd to you. This is called "defensive programming" - you prevent the application from being used in an unexpected way, and instead of relying on the end user to be rational with their choice, we "defend" our function from unexpected inputs.

For example, consider this small app:

```r
library(shiny)
ui <- function(){
  tagList(
    # Designing an interface that lets the
    # user select a species from iris,
    # then display a plot() of this dataset
    selectInput(
      "species",
      "Choose one or more species",
      choices = unique(iris$Species),
      multiple = TRUE,
      selected = unique(iris$Species)[1]
    ),
    plotOutput("plot")
  )
}

server <- function(
  input,
  output,
  session
){
  # Taking the species as input, and returning the plot
  # of the filtered dataset
  output$plot <- renderPlot({
    plot(
      iris[ iris$Species %in% input$species, ]
    )
  })
}

shinyApp(ui, server)
```

What is wrong with this app? Probably nothing from a developer point of view - there is a label stating that one should select one or more elements from the drop-down list, and then something is plotted below. Pretty standard. But what happens if the drop-down is empty? Our first thought would be that this would never happen, as it is explicitly specified that there should be one or more elements selected. In fact, chances are that even with this label, users will eventually end up with an empty selectInput(), leading to the printing of a red error where the plot should be. We are lucky here, as the error only prevents the plot from being displayed; other errors could make the application crash.

What should we do? **Adopt a defensive programming mindset**. Every time you create interactive elements, inputs and outputs, or things the user might interact with, ask yourself: "What if that [crazy thing] happens? How do I handle the case where the minimal viable requirements for my app are not met?" And in fact, you should not be focusing on that only for the user side - the back-end should also be examined for potential unexpected behaviors. For example, if your {shiny} app relies on a database connection, you should gracefully check that the connection is possible, and if it is not, send a message to your user that the database is not reachable, and that they should either restart the app or come back in a few minutes.

In fact, this is a crucial thing when it comes to making your app successful: **you should always fail gracefully and informatively**. That means that even when your R code fails, the whole app should not fail. If the R code fails for some reason, the user should either get nothing back or an informative bug message, not be faced with a grayish version of the application.[2] Note that using external widgets, like the one from the {DT} package (or any other that binds to an external JavaScript library), can make this principle harder to apply: as you have less control over what is happening when using this widget, gracefully handling errors can be tricky. Indeed, {DT} sometimes returns errors that originates from the user's browser, so that has nothing to do with R. In that case, it might be hard to catch this error and gracefully manage it. The only upside of this error is that it does not crash the whole application.

Because of the way {shiny} is designed, a lot of R errors will make the {shiny} app fail completely. If you have not thought about this upfront, that means that a user might use the app for 10 minutes, do a series of specifications, enter parameters and data, only for the app to completely crash at some point. The user has to then restart from scratch, because there is no native way - from there - to restart from where the app has crashed. This is a very important thing to keep in mind when building {shiny} apps: **once the app has failed, there is no easy way to natively get it back to the moment just before it crashed**, meaning that your users might lose a significant amount of the time they have spent configuring the app.

One good practice is to try, as much as possible, to **wrap all server calls in some form of try-catch** pattern. That way, you can, for example, send a notification to the user if the process fails, either using a {shiny} (Chang et al., 2021a) notification function, an external package like {shinyalert} (Attali and Edwards, 2020), or a custom JavaScript alert like notify.js[3]. Here is a pseudo-code pattern for this using the {attempt} (Fay, 2020) package:

---

[2]If you want something different from this grayish screen when {shiny} fails, you can have a look at the {sever} package (Coene, 2021), which allows to implement custom disconnected screen and error messages.

[3]https://github.com/ColinFay/notifyjsexample

```r
library(shiny)
ui <- function(){
  # Here, we would define the interface
  tagList(
    # [...]
  )
}

server <- function(
  input,
  output,
  session
){
  # We are attempting to connect to the database,
  # using a `connect_db()` connection
  conn <- attempt::attempt({
    connect_db()
  })
  # if ever this connection failed, we notify the user
  # about this failed connection, so that they can know
  # what has gone wrong
  if (attempt::is_try_error(conn)){
    # Notify the user
    send_notification("Could not connect")
  } else {
    # Continue computing if the connection was successful
    continue_computing()
  }
}

shinyApp(ui, server)
```

6.1.2 Building a self-evident app (or at least self-explanatory)

One of the goals of a usable app is to make it self-evident, and fall back to a self-explanatory app if the first option is too complex a goal. What is the difference between the two?

- self-evident: "Not needing to be demonstrated or explained; obvious." lexico.com[4]

---

[4] https://www.lexico.com/en/definition/self_evident

- self-explanatory: "Easily understood; not needing explanation." `https://www.lexico.com/en/definition/self_explanatory`

The first is that the app is designed in such a way that there is no learning curve to using it. A self-explanatory app has a small learning curve, but it is designed in a way that will make the user understand it in a matter of seconds.

Let's, for example, get back to our `{tidytuesday201942}` (Fay, 2021k) application available at connect.thinkr.fr/tidytuesday201942[5]. By itself, this application is not self-evident: you need to have a certain amount of of background knowledge before understanding what this application was designed for. For example, you might need to have a vague sense of what `tidytuesday` is. If you do not, you will have to read the home text, which will help you understand what this is. Then, if we have a look at the menu elements, we see that these are a series of functions from `{ggplot2}` (Wickham et al., 2021a): without any background about the package, you might find it difficult to understand what this app actually does.

Yet, if you want to understand what this app is designed for, you will find enough information either on the home page or in the About section, with external links if needed. And of course, when building apps, context matters. The `{tidytuesday201942}` app is one that has been developed in the context of `tidytuesday`, an online weekly event for learning data analysis, mainly through the use of `{tidyverse}` packages. There is a good chance visitors of the app will already know what `{ggplot2}` is when visiting the app.

A. About the "Rule of Least Surprise"

This rule is also known as "Principle of Least Astonishment."

Rule of Least Surprise: In interface design, always do the least surprising thing.

The Art of UNIX Programming (Raymond, 2003)

When we are browsing the web, **we have a series of pre-conceptions about what things are and what they do.** For example, we expect an underline text to be clickable, so there is a good chance that if you use underlined text in your app, the user will try to click on it. Usually, the link is also

[5]https://connect.thinkr.fr/tidytuesday201942/

colored differently from the rest of the text. The same goes for the pointer of the mouse, which usually switches from an arrow to a small hand with a finger up. A lot of other conventions exist on the web, and you should endeavor to follow them: a clickable link should have at least one of the properties we just described—and if it is neither underlined nor colored and does not change the pointer when it is hovered, chances are that the user will not click on it.

Just imagine for a second if our "Download" button in the {tidytuesday201942} app did not actually download the graph you had generated. Even more, imagine if this button did not download the graph but something else. How would you feel about this experience?

And it is not just about links: almost every visual element on a web page is surrounded by conventions. Buttons should have borders. Links should appear clickable. Bigger texts are headers, the bigger the more important. Elements that are "visually nested" are related.

Of course, this is not an absolute rule, and there is always room for creativity when it comes to design, but you should keep in mind that too much surprise can lead to users being lost when it comes to understanding how to use the application.

Weirdly enough, that is an easy thing to spot when we arrive on a web page or an app: it can either feel "natural", or you can immediately see that something is off. The hard thing is that it is something you spot when you are a new-comer: developing the app makes us so familiar with the app that we might miss when something is not used the way it is conventionally used.[6]

Let's exemplify this with the "Render" button from the {tidytuesday201942}[7] application. This app is built on top of Boot-strap 4, which has no CSS class for a {shiny} action button.[8] Result: without any further CSS, the buttons do not come out as buttons, making it harder to decipher that they are actually buttons. Compare the native design shown in Figure 6.3 to the one with a little bit of CSS (which is the one online) shown in Figure 6.4.

Yes, it is subtle, yet the second version of the button is clearer to understand.

Least surprise is crucial to make the user experience a good one: users rarely think that if something is behaving unexpectedly on an app, it is because of the app: they will usually think it is their fault. Same goes for the application failing or behaving in an unexpected way: most users think they are "doing it wrong", instead of blaming the designer of the software.

---

[6]For a good summary of these, see "The cranky user: The Principle of Least Aston-ishmen" https://www.ibm.com/developerworks/web/library/us-cranky10/us-cranky10-pdf.pdf

[7]https://connect.thinkr.fr/tidytuesday201942/

[8]{shiny} is built on top of Bootstrap 3, and the action buttons are of class btn-default, which was removed in Bootstrap 4.

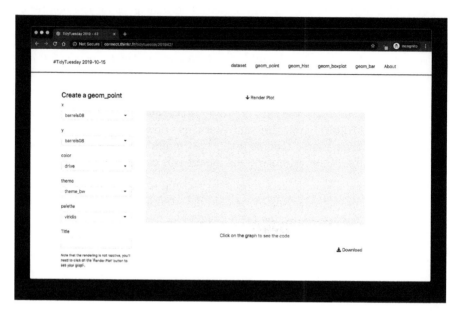

FIGURE 6.3: Snapshot of {tidytuesday201942} without borders around the "Render Plot" button.

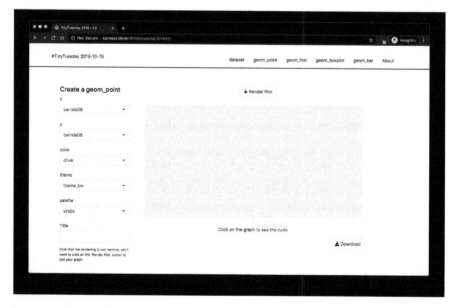

FIGURE 6.4: Snapshot of {tidytuesday201942} with borders around the "Render Plot" button.

> When users are astonished they usually assume that they have made a mistake; they are unlikely to realize that the page has astonished them. They are more likely to feel that they are at fault for not anticipating the page. Don't take advantage of this; making users feel stupid is not endearing.
>
> *The cranky user: The Principle of Least Astonishment* (https://www.ibm.com/developerworks/web/library/us-cranky10/us-cranky10-pdf.pdf)

B. Thinking about progression

If there is a progression in your app, you should **design a clear pattern of moving forward**. If you need to bring your user from step 1 to step 7, you need to guide them through the whole process, and it can be as simple as putting "Next" buttons on the bottom of each page.

Inside your app, this progression has to be clear, even more if step n+1 relies on the inputs from n. A good and simple way to do that is to hide elements at step n+1 until all the requirements are fulfilled at step n. Indeed, you can be sure that if step 2 relies on step 1 and you did not hide step 2 until you have everything you need, users will go to step 2 too soon.

Another way to help this readability is to ensure some kind of linear logic through the app: step 1, data upload, step 2, data cleaning, step 3, data visualization, step 4, exporting the report. And organized your application around this logic, from left to right / right to left, or from top to bottom.

Let's compare {tidytuesday201942} to {hexmake}—one has a clear progression, {hexmake}, and has been designed as such: the upper menus design the stickers, and then once they are filled you can download them. There is a progression here, from top to bottom. On the other hand, {tidytuesday201942} does not have a real progression inside it: you can navigate from one tab to the other at will. Hence there are no visual clues of progression on that app.

C. Inputs and errors

You're the one developing the app, so of course you are conscious of all the inputs that are needed to complete a specific task. But your users might be new to the app; distracted while reading, they might not clearly understand what they are doing, maybe they do not really want to use your app but are forced to by their boss. Or maybe your app is a little bit hard to understand, so it is hard to know what to do at first.

When building your app, you should **make sure that if an input is necessary, it is made clear inside the app that it is**. One way to do this is simply by hiding UI elements that cannot be used until all the necessary inputs are there: for example, if a plot fails at rendering unless you have provided a selection, do not try to render this plot unless the selection is done. If you are building a dashboard and tab 2 needs specific inputs from tab 1, and tab 3 specific inputs from tab 2, then be sure that tabs 2 and 3 are not clickable/available until all the required inputs are filled. That way, you can help the user navigate through the app, by reducing the cognitive load of having to be sure that everything is correctly set up: if it is not clickable, that is because something is missing.

And do this for all the elements in your app: for example, with {hexmake}, we start with filled fields and a hex sticker which is ready, so that even if you start with the download part, the application would still work. If we had chosen another pattern, such as making the user fill in everything before being able to download, we would have needed to make downloading impossible until all fields are filled. Another example from this application is the use of a MongoDB back-end to store the hex stickers: if the application is launched with `with_mongo` set to FALSE, the user will not see any buttons or field that refers to this option.

Think about all the times when you are ordering something on the internet, and need to fill specific fields before being able to click on the "Validate" button. Well, apply that approach to your app; that will prevent unwanted mistakes.

Note that when using the `golem::use_utils_ui()` function, you will end with a script of UI tools, one being `with_red_star`, which adds a little red star at the end of the text you are entering, a common pattern for signifying that a field is mandatory:

```
with_red_star("Enter your name here")
```

Also, be generous when it comes to errors: it is rather frustrating for a user to see an app crash without any explanation about what went wrong. If something fails or behaves unexpectedly, error messages are a key feature to help your user get on the right track. And, at the same time, helping them correct themselves after an error is the best way to save you time answering angry emails!

Let's refactor our app from before, using the {shinyFeedback} (Merlino and Howard, 2020) package.

```r
library(shiny)
library(shinyFeedback)

ui <- function(){
  tagList(
    # Attaching the {shinyFeedback} dependencies
    useShinyFeedback(),
    # Recreating our selectInput + plot from before
    selectInput(
      "species",
      "Choose one or more species",
      choices = unique(iris$Species),
      multiple = TRUE,
      selected = unique(iris$Species)[1]
    ),
    plotOutput("plt")
  )
}

server <- function(
  input,
  output,
  session
){
  output$plt <- renderPlot({
    # If the length of the input is 0
    # (i.e. nothing is selected),we show
    # a feedback to the user in the form of a text
    # If the length > 0, we remove the feedback.
    if (length(input$species) == 0){
      showFeedbackWarning(
        inputId = "species",
        text = "Select at least one Species"
      )
    } else {
      hideFeedback("species")
    }
    # req() allows to stop further code execution
    # if the condition is not a truthy.
    # Hence if input$species is NULL, the computation
    # will be stopped here.
    req(input$species)
    plot(
      iris[ iris$Species %in% input$species, ]
```

```
    )
  })
}
```

```
shinyApp(ui, server)
```

Here, as a user, it is way easier to understand what went wrong: we have moved from a red error `Error: need finite 'xlim' values` to a pop-up explaining what went wrong in the way the user configured the app. Perfect way to reduce your bug tracker incoming tickets!

This is a way to do it natively in `{shiny}`, but note that you can also use the `{shinyAlert}` package to implement alerts. It is also possible to build your own with a little bit of HTML, CSS and JavaScript, as shown in the `notifyjsexample` repository[9].

6.2   The danger of feature-creep

6.2.1   What is feature-creep?

Even more often (at least in the commercial software world) excessive complexity comes from project requirements that are based on the marketing fad of the month rather than the reality of what customers want or software can actually deliver. Many a good design has been smothered under marketing's pile of "checklist features"—features that, often, no customer will ever use. And a vicious circle operates; the competition thinks it has to compete with chrome by adding more chrome. Pretty soon, massive bloat is the industry standard and everyone is using huge, buggy programs not even their developers can love.

The Art of UNIX Programming (Raymond, 2003)

---

[9]https://github.com/ColinFay/notifyjsexample

Feature-creep is the process of **adding features to the app that compli-cate the usage and the maintenance of the product, to the point that extreme feature-creep can lead to the product being entirely unus-able and completely impossible to maintain**. This movement always starts well-intentioned: easier navigation, more information, more visualiza-tions, modifiable elements, and so on and so forth. It can come from project managers or devs, but users can also be responsible for asking for more and more features in the app. If you are working in a context where the app speci-fications were designed by the users, or where you regularly meet the users for their feedback, they will most often be asking for more than what is efficiently implementable. Behind feature-creep, there is always a will to make the user experience better, but adding more and more things most often leads to a slower app, worse user experience, steeper learning curve, and all these bad states that you do not want for your app.

Let's take a rather common data analytic process: querying the data, cleaning it, then plotting and summarizing it. And let's say that we want to add to this a simple admin dashboard that tracks what the users do in the app. It's pretty tempting to think of this as a single entity and throw the whole codebase into one big project and hope for the best. But let's decompose what we have for a minute: one task is querying and cleaning, one other is analyzing, and one other is administration. What is the point of having one big app for these three different tasks? Splitting this project into three smaller apps will keep you from having a large app which is harder to maintain, and that might not perform as well. Indeed, if you put everything into the same app, you will have to add extra mechanisms to prevent the admin panel from loading if your user simply wants to go to the extraction step, and inversely, a user visiting the admin panel probably does not need the extraction and analysis back-end to be loaded when they simply want to browse the way other users have been using the app.

Rule of Parsimony: Write a big program only when it is clear by demonstration that nothing else will do.

The Art of UNIX Programming (Raymond, 2003)

But let's focus on a smaller scope, and think about some things that can be thought of as feature-creeping your {shiny} app.

6.2.2 Too much reactivity

When designing an app, you will be designing the way users will navigate through the app. And most of the time, we design with the idea that the users will perform a "correct selection" pattern. Something like: "The user will select 40 on the `sliderInput()` and the plot will update automatically. Then the user will select the element they need in the `selectInput()` and the plot will update automatically". *When in reality what will happen is:* "The user will click on the slider, aim at 40 but will reach 45, then 37, then 42, before having the right amount of 40. Then they will select something in the `selectInput()`, but chances are, not the correct one from the first time."

In real-life usage, **people make mistakes while using the app** (and even more when discovering the application): they do not move the sliders to the right place, so if the application reacts to all of the moves, the experience using the app can be bad: in the example above, full reactivity means that you will get 4 "wrong" computations of the plot before getting it right.

In the `{tidytuesday201942}` application example, let's imagine that all the elements on the left automatically update the plot: especially in the context of a learning tool, reacting to any configuration change will launch a lot of useless computation, slowing the app in the long run, and making the user experience poorer.

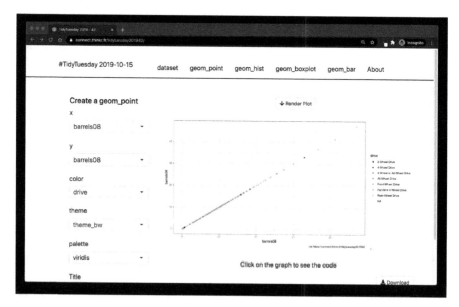

FIGURE 6.5: Snapshot of the `{tidytuesday201942}` `{shiny}` application.

What should we do? Prevent ourselves from implementing "full reactivity": instead, we will add a user input that will launch the computation. The simplest solution iss a button so that the user signals to the application that now they are ready for the application to compute what they have parameterized.

6.2.3 Too much interactivity

Users **love** interactive elements. Maybe too much. If you present a user with a choice between a simple graph and a dynamic one, chances are that they will spontaneously go for the dynamic graph. Yet, dynamic is not always the solution, and for several reasons.

A. Speed

Dynamic elements are slower to render than fixed ones. Most of the time (if not always), rendering dynamic elements means that you will bind some external libraries, and maybe you will have to make R convert data from one format to another. For example, rendering a {ggplot2} plot will be faster than rendering a ggplotly() plot, which has to convert from one format to another.[10]

That being said, not all visualization libraries are created equal, and choosing interactive visualization will not automatically lead to poorer performance: just keep in mind that this can happen.

Finally, if you do choose to use an interactive library for your application, try to, if possible, stick with one: it's easier for you as a developer as it will lower the potential conflicts between libraries, and for the user, who will have to "learn" only one interactive mechanism.

B. Visual noise

More interactivity can lead to an element being less straightforward to understand. Think for a minute about the {plotly} outputs, as seen on Figure 6.6. They are awesome if you need this kind of interactivity, but for a common plot there might be too many things to understand. Instead of focusing on the data, a lot of things show: buttons to zoom, to do selection, to export in png, and things like that. With this kind of graph, users might lose some time focusing on understanding what the buttons do and why they are there, instead of focusing on what matters: getting insights from the data.

Of course, these features are awesome if you need them: exploring data interactively is a fundamental strength for an application when the context is right. But if there is no solid reason for using an interactive table, use a standard

---

[10]Well, maybe the native {plotly} (Sievert, 2020) implementation is faster, but you get the spirit.

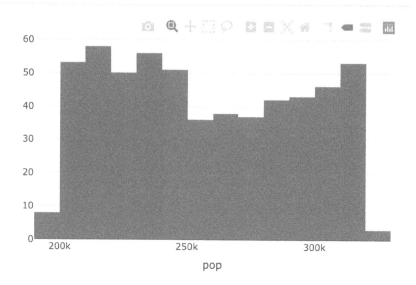

FIGURE 6.6: Output of a {plotly} output, with all available buttons shown.

HTML table. In other words, do not make things interactive if there is no value in adding interactivity; for example, if you have a small table and the users do not need to sort the table, filter, or navigate in pages, `datatable()` from {DT} (Xie et al., 2021) will add more visual noise than adding value to the application.

Adding interactivity widgets (in most cases) means adding visual elements to your original content: in other words, you are adding visual components that might distract the user from focusing on the content of the information.

To sum up, a good rule to live by is that you should not add a feature for the sake of adding a feature.

Less is more.

Ludwig Mies van der Rohe (http://www.masterofdetails.fr/mies-van-der-rohe-less-is-more/)

6.3   Web accessibility

6.3.1   About accessibility

When building professional {shiny} applications, you have to keep in mind that, potentially, this app will be consumed by a large audience. **A large audience means that there is a chance that your app will be used by people with visual, mobility, or maybe cognitive disabilities.**[11] Web accessibility deals with the process of making the web available to people with disabilities.

> The Web is fundamentally designed to work for all people, whatever their hardware, software, language, location, or ability. When the Web meets this goal, it is accessible to people with a diverse range of hearing, movement, sight, and cognitive ability.
>
> *Accessibility in Context - The Web Accessibility Initiative* (https://www.w3.org/WAI/fundamentals/accessibility-intro/)

When learning to code a web app through "canonical" courses, you will be introduced to web accessibility very early. For example, you can learn about this straight from the first chapter of learn.freecodecamp.org[12]. The first course, "Responsive Web Design Certification", has a chapter on web accessibility just after the one on HTML and CSS.

6.3.2   Making your app accessible

A. Hierarchy

Headers are not just there to make your application more stylish. <h1> to <h6> are there so they can create a hierarchy inside your web page: <h1> being more important (hierarchically speaking) than <h2>. In a perfectly designed website, you would only have one header of level 1, a small number of level 2 headers,

---

[11]And of course, other type of disabilities.
[12]https://learn.freecodecamp.org/

more headers of level 3, etc. These elements are used by screen readers (devices used by blind people) to understand how the page is organized.

Hence, you should not rely on the header level for styling: do not use an `<h1>` because you need a larger title somewhere in your app. If you want to increase the size of a header, use CSS, which we will see in an upcoming chapter.

B. HTML element: Semantic tags, and tag metadata

In HTML, there are two kinds of elements: the ones without "meanings" like `<div>` or `<span>`, and the ones which are considered meaningful, like `<title>` or `<article>`. The second ones are called "semantic tags", as they have a specific meaning in the sense that they define what they contain. Same thing as with headers; these elements are crucial for the screen readers to understand what the page contains.

```r
library(htmltools)
# Using the `article` tag for a better semantic
tags$article(
  tags$h2("Title"),
  tags$div("Content")
)
```

One other HTML method you can use is tag attributes as metadata. Tag attributes are complementary elements you can add to a tag to add information: most of the time, you will be using it to add a CSS class, an identifier, or maybe some events like `onclick`.[13] But these can also be used to add, for example, an alternate text to an image: this `alt` being the one which is read when the image is not available, either because the page could not reach the resource, or because the person navigating the app is using a screen-to-speech technology. To do this, we can use the `tagAppendAttributes()` function from {shiny}, which allows us to add attributes to an HTML element.

```r
library(shiny)
library(magrittr)
ui <- function(){
  # Generating a UI with one plot
  tagList(
    plotOutput("plot") %>%
      # Adding the `alt` attribute to our plot
```

---

[13]See the JavaScript chapter.

```
        tagAppendAttributes(alt = "Plot of iris")
  )
}

server <- function(
  input,
  output,
  session
){
  # Generating the plot from the server side,
  # no modification here
  output$plot <- renderPlot({
    plot(iris)
  })
}

shinyApp(ui, server)
```

What makes these two things similar (semantic tags and tag metadata) is that they are both unseen by users without any impairment: if the image is correctly rendered and the user is capable of reading images, chances are that this user will see the image. But these elements are made for people with disabilities, and especially users who might be using screen-to-speech technologies: these visitors use a software that scans the textual content of the page and reads it, and that helps navigate through the page.

This navigation is also crucial when it comes to screen-to-speech technology: such software will be able to read the `<title>` tag, jump to the `<nav>`, or straight to the `<article>` on the page. Hence the importance of structuring the page: these technologies need the app to be built in a structured way, so that it is possible to jump from one section to another, and other common tasks a fully capable user will commonly do.

Some other tags exist and can be used for semantic purpose: for example, `<address>`, `<video>`, or `<label>`.

C. Navigation

Your app user might also have mobility impairment. For example, some with Parkinson's disease might be using your app, or someone with a handicap making it harder for them to move their hand and click. For these users, moving an arm to grab the mouse might be challenging, and they might be navigating the web using their keyboard only.

When building your app, thinking about how these users will be able to use it is crucial: maybe there are so many buttons to which they need to **move their mouse and eventually click** that they will not be able to use it. As much as possible, make everything doable with a keyboard: for example, if you have a `textInput()` with a validation button below, allow the user to validate by pressing ENTER on their keyboard. This can, for example, be done with the {nter} package, which is available only on GitHub[14] at the time of writing these lines.

```r
# Adapted from https://github.com/JohnCoene/nter
library(nter)
library(shiny)

ui <- fluidPage(
  # Setting a text input and a button
  textInput("text", ""),
  # This button will be clicked when 'Enter' is pressed in
  # the textInput text
  actionButton("send", "Do not click hit enter"),
  verbatimTextOutput("typed"),
  # define the rule
  nter("send", "text")
)

server <- function(input, output) {

  r <- reactiveValues()

  # Define the behavior on click
  observeEvent( input$send , {
    r$printed <- input$text
  })

  # Render the text
  output$typed <- renderPrint({
    r$printed
  })
}

shinyApp(ui, server)
```

---

[14]https://github.com/JohnCoene/nter

D. Color choices

Color blindness is also a common impairment when it comes to web accessibility. And it is a rather common deficiency: according to colourblindawareness.org[15], "color (color) blindness (color vision deficiency, or CVD) affects approximately 1 in 12 men (8%) and 1 in 200 women in the world".

Keeping in mind this prevalence of color blindness is even more important in the context of {shiny}, where we are developing data science products, which most often include data visualization. If designed wrong, dataviz can be unreadable for some specific type of color blindness. That is why we recommend using the viridis (Garnier, 2021) palette, which has been created to be readable by the most common types of color blindness.

Here are, for example, a visualization through the lens of various typed of color blindness:

```r
# This function generates a plot for an
# internal matrix, and takes a palette as
# parameter so that we can display the
# plot using various palettes, as the
# palette should be a function
with_palette <- function(palette) {
  x <- y <- seq(-8 * pi, 8 * pi, len = 40)
  r <- sqrt(outer(x^2, y^2, "+"))
  z <- cos(r^2) * exp(-r / (2 * pi))
  filled.contour(
    z,
    axes = FALSE,
    color.palette = palette,
    asp = 1
  )
}
```

With the jet.colors palette from {matlab} (Roebuck, 2014)

```r
with_palette(matlab::jet.colors)
```

See Figure 6.7.

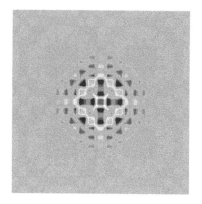

FIGURE 6.7: Original view of `jet.colors` palette from `{matlab}`.

```
with_palette(viridis::viridis)
```

See Figure 6.8.

Even without color blindness, it's already way more readable. But let's now use the `{dichromat}` (Lumley, 2013) package to simulate color blindness.

```
library(dichromat)
```

- Simulation of deuteranopia with `jet.colors` and `viridis`

```
deutan_jet_color <- function(n){
  cols <- matlab::jet.colors(n)
  dichromat(cols, type = "deutan")
}
with_palette( deutan_jet_color )
```

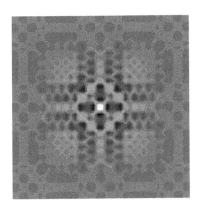

FIGURE 6.8: Original view of `viridis` palette from {viridis}.

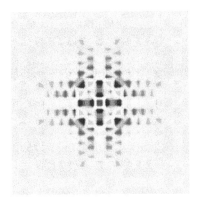

FIGURE 6.9: View of `jet.colors` palette for a deuteranopian.

See Figure 6.9.

```
deutan_viridis <- function(n){
  cols <- viridis::viridis(n)
  dichromat(cols, type = "deutan")
}
with_palette( deutan_viridis )
```

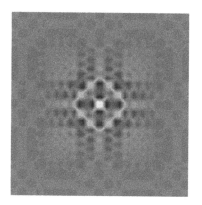 

FIGURE 6.10: View of `viridis` palette for a deuteranopian.

See Figure 6.10.

- Simulation of protanopia with `jet.colors` and `viridis`

```
protan_jet_color <- function(n){
  cols <- matlab::jet.colors(n)
  dichromat(cols, type = "protan")
}
with_palette( protan_jet_color )
```

FIGURE 6.11: View of `jet.colors` palette for a protanopian.

See Figure 6.11.

```
protan_viridis <- function(n){
  cols <- viridis::viridis(n)
  dichromat(cols, type = "protan")
}
with_palette( protan_viridis )
```

See Figure 6.12.

- Simulation of tritanopia with `jet.colors` and `viridis`

```
tritan_jet_color <- function(n){
  cols <- matlab::jet.colors(n)
  dichromat(cols, type = "tritan")
}
with_palette( tritan_jet_color )
```

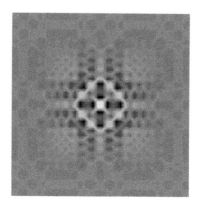

FIGURE 6.12: View of `viridis` palette for a protanopian.

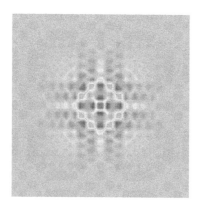
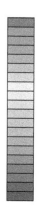

FIGURE 6.13: View of `jet.colors` palette for a tritanopian.

See Figure 6.13.

```
tritan_viridis <- function(n){
  cols <- viridis::viridis(n)
  dichromat(cols, type = "tritan")
}
with_palette( tritan_viridis )
```

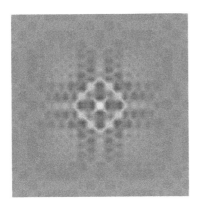

FIGURE 6.14: View of `viridis` palette for a tritanopian.

See Figure 6.14.

As you can see, the `viridis` palette always gives a more readable graph than the `jet.colors` one. And, on the plus side, it looks fantastic. Do not hesitate to try and use it!

6.3.3 Evaluating your app accessibility and further reading

A. Emulate vision deficiency using Google Chrome

Google Chrome has a built-in feature that allows you to simulate some vision deficiency. To access this feature, open your developer console, then open the

"More Tools" > "Rendering" menu. There, you will find at the very bottom an input called "Emulate vision deficiencies", which will allow you to simulate Blurred vision, and four types of color blindness.

For example, Figure 6.15 and Figure 6.16 emulate blurred vision or deuteranopia on the {hexmake} app.

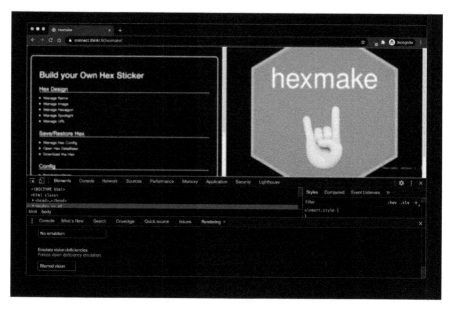

FIGURE 6.15: Emulating blurred vision with Google Chrome.

B. External tools

There are several tools on the web that can evaluate the accessibility of your web page. You can also use a Google Chrome built-in tool called `Lighthouse` (we will come back to it in the Testing chapter).

- IBM Equal Access Toolkit[16] is an open source tool to monitor the accessibility of a web application and comes with Google Chrome and Firefox Extensions.

- Evaluating Web Accessibility[17] comes with lengthy reports and advice about checking the accessibility of your website.

- `https://www.webaccessibility.com/` has an online checker for web page accessibility, and allows you to freely test 5 pages. The result of a test on the {hexmake} application can be seen on Figure 6.17.

---

[16]`https://github.com/IBMa/equal-access`
[17]`https://www.w3.org/WAI/test-evaluate/`

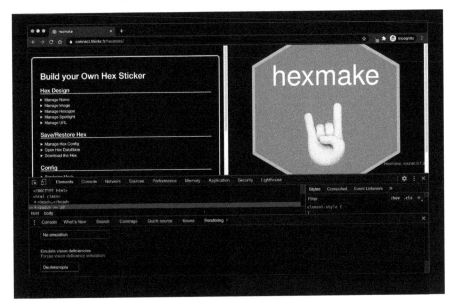

FIGURE 6.16: Emulating deuteranopia with Google Chrome.

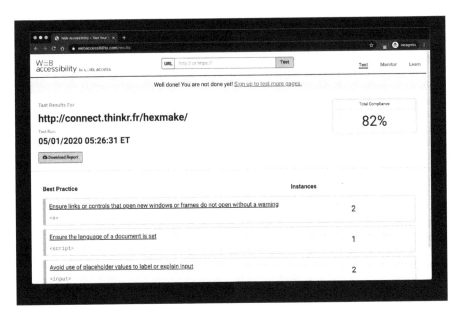

FIGURE 6.17: Web accessibility results for the {hexmake} application.

Note that you can also add a Chrome or Firefox extension for https:
//www.webaccessibility.com, making it more straightforward to run your
accessibility tests. It also comes with tools for Java and JavaScript, and no-

tably with a NodeJS module, so it can be used programmatically, for example, in your Continuous Integration suite.

7

Don't Rush into Coding

7.1 Designing before coding

> You have to believe that software design is a craft worth all the intelligence, creativity, and passion you can muster. Otherwise you will not look past the easy, stereotyped ways of approaching design and implementation; **you will rush into coding when you should be thinking**. You'll carelessly complicate when you should be relentlessly simplifying—and you'll wonder why your code bloats and debugging is so hard.
>
> *The Art of UNIX Programming* (Raymond, 2003) (Our bold.)

7.1.1 The urge to code

At the moment you receive the specifications for your app, it is tempting to rush into coding. And that is perfectly normal: we're R developers because we love building software, so as soon as a problem emerges, our brain starts thinking about technical implementation, packages, pieces of code, and all these things that we love to do when we are building an application.

But **rushing into coding from the very beginning is not the safest way to go**. Focusing on technical details from the very beginning can make you miss the big picture, be it for the whole app if you are in charge of the project, or for the piece of the whole app that you have been assigned. Have you ever faced a situation in a coding project where you tell yourself "Oh, I wish I had realized this sooner, because now I need to refactor a lot of my code for this specific thing"? Yes, we all have been in this situation: realizing too late that the thing we have implemented does not work with another feature we discover along the road. And what about "Oh I wish I had realized sooner that this package existed before trying to implement my own functions to do that!"[1] Same thing: we're jumping straight into solving a programming problem when someone else has open-sourced a solution to this very same problem.

Of course, implementing your own solution might be a good thing in specific cases: avoiding heavy dependencies, incompatible licensing, the joy of the intellectual challenge, but **when building production software, it is safer to go for an existing solution if there is one and it fits in the project: existing packages/software that are widely used by the community and by the industry benefit from wider testing, wider documentation, and a larger audience if you need to ask questions.** And of course, it saves time, be it immediately or in the long run: re-using an existing solution allows you to save time re-implementing it, so you save time today, but it also prevents you from having to detect and correct bugs, saving you time tomorrow.[2]

Note also that assessing that a dependency/technology is a good choice for an application is not an easy task: there is a difference between *thinking* something will be the good choice and *knowing* that this choice is the correct one. Most of the time, when faced with a new technology, it makes sense to take some time to write a small prototype that tests the features we want to use. This process of prototyping small applications to test features is made easier notably by using the {shinipsum} package, which we will see in Chapter 9.

Before rushing into coding, take some time to conceptualize your application/modules on a piece of paper. That will help you get the big picture of the piece of code you will be writing: what are the inputs, what are the outputs, what packages/services can you use inside your application, how will it fit in the rest of the project, and so on and so forth.

7.1.2   Knowing where to search

Being a good developer is knowing where to search, and what to search for. Here is a non-exhaustive list of places you can look if you are stuck/looking for existing packages.

---

[1] Given the dynamic of the R community, there is no way to completely avoid this: new packages are created and publish every day, so there is no way to be aware of everything. But trying to assess what exists before jumping into coding will definitely save you some time in the long run.

[2] Of course, it is not an absolute rule: you might also inherit from the bug created by the open source solution.

R and {shiny}

- CRAN Task View: Web Technologies and Services[3] and CRAN Task View: Databases with R[4], which will be useful for interacting with web technologies and databases.
- The cloudyr project[5], which focuses on cloud services and R.
- METACRAN[6], which is a search engine for R packages.
- GitHub search using `language:R`[7]: When doing a search on GitHub, do not forget to add the language-specific tag.
- RStudio Community[8] has a series of posts about {shiny}: questions, announcements, best practices, etc.

Web

- Mozilla developer center[9] is one of the most comprehensive resource platforms when it comes to web technologies (HTML, CSS, and JavaScript)
- Google Developer Center[10] also has a series of resources that can be helpful when it comes to web technologies.
- FreeCodeCamp[11] contains more than 2000 hours of free courses about web technologies, plus a blog and forum.

7.1.3 About concept map

Using a concept map to think about your app can be a valuable method to help you grasp the big picture of your application.

Concept maps are a widely used tool, in the software engineering world and in many other fields. The idea with concept maps is to take a piece of paper (or a digital tool) and **draw all the concepts that come to mind for a specific topic, and all the relationships that link these concepts together**. Drawing a concept map is a way to organize the knowledge of a specific topic.

When doing this for a piece of software, we are not trying to add technical details about the way things are implemented: we are listing the various "actors" (the concepts) around our app, with the relationships they have. For example, Figure 7.1 is a very simple concept map of the {hexmake} (Fay, 2021f) app.

---

[3]https://cran.r-project.org/web/views/WebTechnologies.html
[4]https://cran.r-project.org/web/views/Databases.html
[5]https://cloudyr.github.io/
[6]https://r-pkg.org/
[7]https://github.com/search?q=language%3AR
[8]https://community.rstudio.com/c/shiny/8
[9]https://developer.mozilla.org/
[10]https://developers.google.com/
[11]https://www.freecodecamp.org/

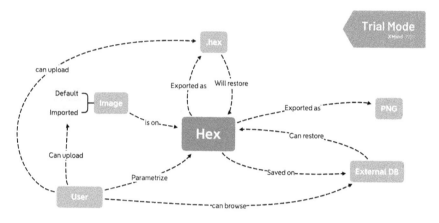

FIGURE 7.1: {hexmake} concept map, built with XMind (https://www.xmind.net).

As you can see, we are not detailing the technical implementations: we are not writing the external database specification, the connection process, how the different modules interact with each other, etc. The goal of a concept map is to think about the big picture, to see the "who and what" of the application. Here, creating this concept map helps us list the flow of the app: there is a user that wants to configure a hex, built with a default image or with an uploaded one, and once this hex is finished, the user can either download it or register it in a database. This database can be browsed and restore hex. The user can also export a .hex file, that can restore an app configuration.

Once this general flow is written down, you can get back to it several times during the process of building the app, but it is also a perfect tool at the end to see if everything is in place: once the application is finished, we can question it:

- Can we point to any concept and confirm it's there?
- Can we look at every relationship and see they all work as expected?

Deciding which level of detail you want to put in your concept map depends; "simple" applications probably do not need complex maps. And that also depends on how precise the specifications are, and how many people are working on the project: the concept map is a valuable tool when it comes to communication, as it allows people involved in the project to have visual clues of the conceptual architecture of the application.

But beware: very complex maps are also unreadable! In that case, it might make sense to divide into several concept maps: one with the "big picture", and smaller ones that focus on specific components of your application.

7.2 Ask questions

Before starting to code, the safe call will be to ask your team/client (depending on the project) a series of questions just to get a good grasp of the whole project.

Here is a (non-exhaustive) list of information you might need along the way.

Side note: Of course, these questions do not cover the core features of the application. We're pretty sure you have thought about covering this already. These are more contextual questions which are not directly linked to the application itself, yet that can be useful down the line.

7.2.1   About the end users

Some questions you might ask:

- Who are the end users of your app?
- Are they tech-literate?
- In which context will they be using your app?
- On what machines (computer, tablet, smartphone, or any other device)?
- Are there any restrictions when it comes to the browser they are using? (For example, are they still using an old version of Internet Explorer?)
- Will they be using the app in their office, on their phone while driving a tractor, in a plant, or while wearing a lab coat?

Those might seem like weird questions if you are just focusing on the very technical side of the app implementation, but think about where the app will be used: the application used while driving agricultural machines might need fewer interactive things, bigger fonts, simpler interface, fewer details, and more direct information. If you are building a {shiny} app for a team of sellers who are always on the road, chances are they will need an app that they can browse from their mobile. And developing for mobiles requires a different kind of mindset.[12]

Another good reason why talking to the users is an important step, is that most of the time, **people writing specifications are not the end users and will either request too many features or not enough**. Do the users really need that much interactive plots? Do they actually need that

---

[12]For developing an app that is mobile first, you can have a look at the great {shinyMobile} (Granjon et al., 2021) package made by the amazing Rinterface (https://rinterface.com/) team.

much granularity in the information? Will they really see a `datatable` of 15k lines? Do they really care about being able to zoom in the `dygraph` so that they can see the point at a minute scale? To what extent does the app have to be fast?

Asking these questions is important, because building interactive widgets makes the app a little bit slower, and shoving in a series of unnecessary widgets will make the user experience worse, adding more cognitive load than necessary. The speed of execution of your app is also an important parameter for your application: getting a sense about the need for speed in your application will allow you to judge whether or not you will have to focus on optimizing code execution.

On top of that, remember all these things we saw in the last chapter about accessibility: some of your end users might have specific accessibility requirements.

7.2.2 Building personas

The persona is a concept borrowed from design and marketing that refers to fictional characters that will serve as a user type. In other words, **a persona is a character that represents the "typical" behavior and traits for a group of users that will interact with your product.**

A persona consists of a description of a fictional person who represents an important customer or user group for the product, and typically presents information about demographics, behavior, product usage, and product-related goals, tasks, attitudes, etc.

Quantitative Evaluation of Personas as Information (Chapman et al., 2008)

Using personas during the design process helps you center your focus on the end user, so that you know who you are creating the application for. Then, while building your application, you can think about how each persona will interact with a given feature: Will they use it? Will they understand it? Do we need to add extra information? Will they find this useful?

Asking these kinds of questions helps you take a step back from feature implementation and re-focus on what matters: we are building application for someone else, who will eventually use it.

The benefits of personas are that they enable designers to envision the end user's needs and wants, remind designers that their own needs are not necessarily the end users' needs, and provide an effective communication tool, which facilitates better design decisions.

Creating and Using Personas in Software Development: Experiences from Practice (Billestrup et al., 2014)

The building of these personas is made easier once you have interacted with the end users, as we suggested in the previous section. Given the answers to these questions, you will be able to draw some common characteristics about the future users of your application.

And don't hesitate to detail these fictional characters as "[p]ersonas are considered to be most useful if they are developed as whole characters, described with enough detail for designers and developers to get a feeling of its personality". (Billestrup et al., 2014)

7.2.3 Pre-existing code-base

From time to time, you are building a {shiny} app on top of an existing code-base: either scripts with business logic, a package if you are lucky, or a PoC for a {shiny} app.

These kinds of projects are often referred to as "brownfield projects", in opposition to "greenfield projects", borrowing the terminology from urban planning: **a greenfield project being one where you are building on "evergreen" lands, while a brownfield project is building on lands that were, for example, industrial lands, and which will need to be sanitized, as they potentially contain waste or pollution, constructions need to be destroyed, roads needs to be deviated, and all these things that can make the urban planning process more complex.** Then, you can extend this to software engineering, where a greenfield project is the one that you start from scratch, and a brownfield project is one where you need to build on top of an existing code-base, implying that you will need to do some extra work before actually working on the project.

> When transforming brownfield projects, we may face significant impediments and problems, especially when no automated testing exists, or when there is a tightly-coupled architecture that prevents small teams from developing, testing, and deploying code independently.
>
> *The DevOps Handbook* (Kim, 2016)

Depending on how you chose to handle it, starting from a codebase that is already written can either be very much helping, or you can be shooting yourself in the foot. Most of the time, {shiny} projects are not built as reproducible infrastructures: you will find a series of library() calls, no functions structure *per se*, no documentation, and no tests. In that case, we would advise you to do it "the hard way", or at least what seems to be the hard way: throw the app away and start from scratch.

Well, not really from scratch: **extract the core business logic of the app and make it a package**. Take some time with the developer(s) that built the current app, so that you can make them extract the core business logic, i.e. all the pieces of code that do not need a reactive context to run. Write documentation for this package, work on testing, and once you are done, call it a day: you now have solid ground for building the back-end, and it is built outside of any reactivity, is not linked to any application, and most of the time it can be used outside of the app. It might actually be more useful than you think: it can serve analysts and data scientists that will benefit from these functions outside of the application, as they can use the business logic functions that are now packaged, and so reusable.

Existing {shiny} projects, in most cases, have not been built by software engineers or web developers—they have been built by data analysts/scientists who wanted to create an interactive PoC for their work. The good news, then, is that you can expect the core algorithms to be pretty solid and innovative. But web development is not their strength: and that is perfectly normal, as it is not their core job. What that implies is that most {shiny} PoCs take shortcuts and rely on hacks, especially when it comes to managing reactivity, which is a beautiful concept for small projects but can be very complex to scale if you are not a software engineer by training; even more, given that R is by nature sequential.

That's why it is better to split the business and app logic from the very beginning (as we have explained in chapter 3): it simplifies the process of refactoring a {shiny} PoC into a production-grade {shiny} application.

7.2.4 Deployment

There are so many considerations about deployment that it will be very hard to list them all, but keep in mind that **if you do not ask questions about where your application will be deployed from the very beginning, sending it to production might become a painful experience**. Of course, it is more or less solved if you are deploying with Docker: if it works in a container on your machine, it should work in production, but it is not as simple as that: for example, building a {shiny} application that will be used by 10 people is not the same as building an application that needs to scale to 50.000 users. Learning at the end of the project that "now we need to scale to a very large user base" might prevent the deployment from being successful, as this kind of scale implies specific consideration while building.

But that is just the tip of the iceberg of things that can happen. Let's stop for a little story: once upon a time, a team of developers was missioned to build an app, and one feature of the app was to do some API requests. So far so good, nothing too complicated, until they discovered that the server where the app was going to be deployed does not have access to the internet, making it impossible to issue API requests from the server. Here, the containers worked on the dev machines, as they had access to the internet. Once deployed, the app stopped working, and the team lost a couple of days of exchanges with the client, trying to debug the API calls, until we realized that the issue was not with the app, but with the production server itself: and nobody in the team, not the developers or the client, thought about asking about internet access for the server.

It's even more important to think about the IT side of your application, as the people writing specs and interacting with you might come from the Data Science team, and they might or might not have discussed with the IT team about deploying the app. There is a chance that they do not have in mind all of what is needed to deploy a {shiny} app on their company server.

For example, maybe your application has a database back-end. For that, you will need to have access to this database, the correct port should be set, and the permission given to the process that executes the {shiny} app to read, and maybe write, to the database. But, **and for good reason**, database managers do not issue read and write permissions to a database without having examined what the app wants to read, and how and where it will write. To sum up, if you do not want to have weeks of delay for your app deployment, start the discussion from the very beginning of the project. That way, even if the process of getting permission to write on the company database takes time, you might have it by the end of the coding marathon.

Part III

Step 2: Prototype

8

Before starting to prototype and build anything, initialize a {golem} (Fay et al., 2021) project! This will help you start your application on solid ground, and once the project is ready to be filled, you can start prototyping right inside it.

The general workflow for "prototype and build" is the following: the project manager sets up a {golem} project, where the first steps are filled, the general structure (potentially with {shiny} module) is set, and then the project is registered to the version control system. Once we have this structure, package and modules combined, we can start prototyping the UI inside the module, work on the CSS and JavaScript elements that might be needed, and the back-end functionalities inside Rmarkdown files. And then, once these two prototyping sides are finished, we work on the integration of everything inside the reactive context.

In this chapter and in chapter 11, we will be presenting the {golem} package in more depth. {golem} is a framework that standardizes the process of building production-ready {shiny} applications.

8.1 Create a {golem}

Once {golem} is installed and available on your computer, you can go to File > New Project… in RStudio, and choose "Package for {shiny} app Using golem" input.

If you want to do it through the command line, you can use:

```
# Creating a golem project from the command line
golem::create_golem(path = "path/to/package")
```

Once you have that, a new project will be launched. Here is the structure of this project:

```
# This is what a default {golem} project looks like
# Listing the files from the `golex` project using {fs}
fs::dir_tree("golex")
```

```
golex
+-- DESCRIPTION
+-- NAMESPACE
+-- R
|   +-- app_config.R
|   +-- app_server.R
|   +-- app_ui.R
|   \-- run_app.R
+-- dev
|   +-- 01_start.R
|   +-- 02_dev.R
|   +-- 03_deploy.R
|   \-- run_dev.R
+-- golex.Rproj
+-- inst
|   +-- app
|   |   \-- www
|   |       \-- favicon.ico
|   \-- golem-config.yml
\-- man
    \-- run_app.Rd
```

If you already have some experience with R packages, most of these files will
appear very familiar to you. That's because a {golem} app IS a package, so it
uses the standard R package structure (and yes, the good news is that everything you know about R packages will work in a {golem}-based application).

8.2   Setting things up with dev/01_start.R

Once you have created your project, the first file that opens is
dev/01_start.R. This file contains a series of commands to run once, at
the start of the project.

8.2.1 Fill the DESCRIPTION and set options

First, fill the DESCRIPTION file by adding information about the package that will contain your app:

```
golem::fill_desc(
  # The Name of the package containing the App
  pkg_name = "ipsumapp",
  # The Title of the package containing the App
  pkg_title = "PKG_TITLE",
  # The Description of the package containing the App
  pkg_description = "PKG_DESC.",
  # Your First Name
  author_first_name = "AUTHOR_FIRST",
  # Your Last Name
  author_last_name = "AUTHOR_LAST",
  # Your Email
  author_email = "AUTHOR@MAIL.COM",
  # The URL of the GitHub Repo (optional)
  repo_url = NULL
)
```

Then, call the `golem::set_golem_options()` function, which will add information to the `golem-config.yml` file, and set the {here} (Müller, 2020) package root sentinel. {here} is an R package designed to handle directory management in R. When used in combination with {golem}, {here} helps ensure that everything you do in your console is performed relatively to the root directory of your project: the one containing the DESCRIPTION of your application. That way, even if you change the working directory of your R session to a subfolder, you will still be able to create modules and CSS files in the correct folder.

8.2.2 Set common files

If you want to use the MIT license, add README, a code of conduct, a lifecycle badge, and NEWS.

```
# You can set another license here
usethis::use_mit_license( name = "Golem User" )
# Add a README, Code of Conduct, lifecycle badge and NEWS.md
# file to your application
```

```
usethis::use_readme_rmd( open = FALSE )
usethis::use_code_of_conduct()
usethis::use_lifecycle_badge( "Experimental" )
usethis::use_news_md( open = FALSE )
```

It's also where you will be invited to use `Git`:

```
usethis::use_git()
```

8.2.3   Use recommended elements

`golem::use_recommended_tests()` and `golem::use_recommended_deps()` sets a default testing infrastructure and adds dependencies to the application.

8.2.4   Add utility functions

These two functions add a file with various functions that can be used along the process of building your app.

See each file in detail for a description of the functions.

```
# These files will create R/golem_utils_ui.R
# and R/golem_utils_server.R
golem::use_utils_ui()
golem::use_utils_server()
```

In this file, you will, for example, find `list_to_li()`, which is a function to turn an R list into an HTML list or `with_red_star()`, a function to add a small red star after a UI input, useful for communicating that an input is mandatory.

8.2.5   Changing the favicon

Favicons are the small icons located on the tab of your browser: in the default application, this favicon is the {golem} hex.

If you want to change the default favicon:

```
golem::use_favicon( path = "path/to/favicon")
```

You're now set! You've successfully initiated the project and can go to dev/02_dev.R.

8.3   Setting infrastructure for prototyping

8.3.1   Add modules in dev/02_dev.R

The golem::add_module() function creates a module in the R folder. The file and the modules will be named after the **name** parameter, by adding mod_ to the R file, and mod_*_ui and mod_*_server to the UI and server functions.

```
# Creating a module skeleton
golem::add_module(name = "my_first_module")
```

```
v File already exists.
* Go to R/mod_my_first_module.R
```

The new file will contain:

```
#' my_first_module UI Function
#'
#' @description A shiny Module.
#'
#' @param id,input,output,session Internal parameters for {shiny}.
#'
#' @noRd
#'
#' @importFrom shiny NS tagList
mod_my_first_module_ui <- function(id){
  ns <- NS(id)
  tagList(

  )
}
```

```
#' my_first_module Server Functions
#'
#' @noRd
mod_my_first_module_server <- function(id){
  moduleServer( id, function(input, output, session){
    ns <- session$ns

  })
}


## To be copied in the UI
# mod_my_first_module_ui("my_first_module_ui_1")

## To be copied in the server
# mod_my_first_module_server("my_first_module_ui_1")
```

Note that to avoid making errors when putting these into your app, the end of the file will contain code that has to be copied and pasted inside your UI and server functions.

This is where you will be adding the core of your app. The first time, these modules will contain prototyped UI for the application, and once the application is ready to be integrated, you will add the core logic here.

8.3.2 Add CSS and JS files

Adding some infrastructure for JavaScript and CSS files from the very beginning can also formalize the set-up: you are giving the rest of your team a specific file where they can write the JavaScript and CSS code.

```
golem::add_js_file( "script" )
```

will generate the following file:

```
$( document ).ready(function() {

});
```

Here, you will have an infrastructure for launching JavaScript code once the application is ready (this code is standard **jQuery** format: we will be back to JavaScript at the end of this book).

```
golem::add_js_handler( "handlers" )
```

will generate the following file:

```
$( document ).ready(function() {
  Shiny.addCustomMessageHandler('fun', function(arg) {

  })
});
```

As you can see, there is already a skeleton for building {shiny} JavaScript handlers.

```
golem::add_css_file( "custom" )
```

will create a blank CSS file inside the inst/app/www folder.

Note that as you are building your application with {golem}, these files will be linked automatically to your application.

9

Building an "ipsum-app"

9.1 Prototyping is crucial

9.1.1 Prototype, then polish

> Prototyping first may help keep you from investing far too much time for marginal gains.
>
> *The Art of UNIX Programming* (Raymond, 2003)

And yet another rule from *The Art of Unix Programming*: "Rule of Optimization: Prototype before polishing. **Get it working before you optimize it.**" Getting things to work before trying to optimize the app is always a good approach:

- **Making things work before working on low-level optimization makes the whole engineering process easier**: having a "minimal viable product" that works, even if slowly and not perfectly, gives a stronger sense of success to the project. For example if you are building a vehicle, it feels more of a success to start with a skateboard than with a wheel: you quickly have a product that can be used to move, not waiting for the end of the project before finally having something useful. Building a skateboard helps the developer maintain a sense of accomplishment throughout the life of the project: the quicker you can have a running program, a MVP (Minimum Viable Product, as seen on Figure 9.1), the better.

One of the really nice things about running your program frequently is that you get to see it running, which is fun, ans that's what programming is all about.

The Unicorn Project (Kim, 2019)

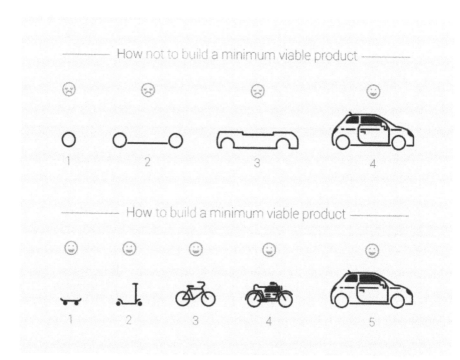

FIGURE 9.1: Building a minimum viable product (MVP).

- **Abstraction is hard, and makes the codebase harder to work with**. You have heard a lot that if you are copying and pasting something more than twice, you should write a function. And with {shiny}, if you are writing a piece of the app more than twice, you should write modules. But while these kinds of abstractions are elegant and optimized, they can make the software harder to work on while building it. So before focusing on turning something into a function, make it work first. As said in *R for Data Science* (Wickham and Grolemund, 2017) about abstraction with {purrr} (Henry and Wickham, 2020):

Once you master these functions, you'll find it takes much less time to solve iteration problems. But you should never feel bad about using a for loop instead of a map function. The map functions are a step up a tower of abstraction, and it can take a long time to get your head around how they work. The important thing is that you solve the problem that you're working on, not write the most concise and elegant code (although that's definitely something you want to strive towards!).

R for Data Science - 21.5 The map functions (Wickham and Grolemund, 2017)

As a small example, we can refer to the binding module from {hexmake} (Fay, 2021f): this module manipulates namespaces, inputs, and session to automatically bind inputs to the R6 object containing the image (see implementation here[1], here[2] and here[3]). That's an elegant solution: instead of duplicating content, we use functions to automatically bind events. But that is a higher level of abstraction: we manipulate different levels of namespacing and inputs, making it harder to reason about when you have to change the codebase.

- It's hard to identify upfront the real bottlenecks of the app. As long as the app is not in a working state, it is very hard to identify the real pieces of code that need to be optimized. Chances are that if you ask yourself upfront what the app bottlenecks will be, you will not aim right. Instead of losing time focusing on specific pieces of code you think need to be optimized, start by having something that works, then optimize

---

[1]https://github.com/ColinFay/hexmake/blob/master/R/mod_binder.R#L29
[2]https://github.com/ColinFay/hexmake/blob/master/R/utils_server.R#L2
[3]https://github.com/ColinFay/hexmake/blob/master/R/mod_left.R#L60

the code. In other words, "Make It Work. Make It Right. Make It Fast", (KentBeck[4]).

- It's easier to spot mistakes when you have something that can run. If a piece of software runs, it is straightforward to check if a change in the codebase breaks the software or not: it either still runs or not.

9.1.2 The "UI first" approach

Using what can be called a "UI first" approach when building an app is in most cases the safest way to go. And for two main reasons.

A. Agreeing on specifications

First of all, it **helps everybody involved in the application to agree on what the app is supposed to do, and once the UI is set, there should be no "surprise implementation"**. Well, at least, this is the best way to reduce the number of changes in the app, as the sooner we have a global idea of the app, the better. It is hard to implement a core new feature once the app is 90% finished, while it would have been way easier to implement it if it had been detected from the very start. Indeed, implementing core features once the app is very advanced can be critical, as our application might not have been thought to work the way it now needs to work, so adding certain elements might lead to a need for change in the core architecture of the app. Once we agree on what elements compose the app, there should be no sudden "oh, the app needs to do that thing now, sorry I hadn't realized that before".

We cannot blame the person ordering the app for not realizing everything needed to build the app: it is really hard to have a mental model of the whole software when we are writing specifications, not to mention when reading them. On the other hand, having a mock application with the UI really helps us realize what the app is doing and how it works, and to agree with the developer that this is actually what we want our application to do (or realize that this is not something we actually need).

Prototyping the UI first should require the least possible computation from the server side of your application. You focus on the appearance of the app: buttons, figures, tables, and graphs, and how they interact with each other. **At that stage of the design process, you will not be focusing on the correctness of the results or graphs: you will be placing elements on the front-end so that you can be sure that everything is there, even if some buttons do not trigger anything**. At that point, the idea is to get the people who are ordering the app to think about what they actually

---

[4]https://wiki.c2.com/?MakeItWorkMakeItRightMakeItFast

need, and there might be a question like "oh, where is the button to download the results in a pdf?". And at that precise moment is the perfect time for a change in specification.

B. Organizing work

A pre-defined UI allows every person involved in the coding process to know which part of the app they are working on, and to be sure that you do not forget anything. As you might be working on the app as a team, you will need to find a strategy for efficiently splitting the work among coders. **It's much easier to work on a piece of the app you can visually identify and integrate in a complete app scenario.** In other words, it is easier to be told "you will be working on the 'Summary' panel from that mock UI" than "you will be working on bullet points 45 to 78 of the specifications".

9.2 Prototyping {shiny}

In the next section, you will be introduced to two packages that can be used when prototyping a user interface: {shinipsum} (Fay and Rochette, 2021b) and {fakir} (Fay and Rochette, 2021a).

9.2.1 Fast UI prototyping with {shinipsum}

When prototyping the UI for an application, we will not be focusing on building the actual computation: **what we need is to create a draft with visual components, so that we can have visual clues about the end result**. To do that, you can use the {shinipsum} package, which has been designed to generate random {shiny} (Chang et al., 2021a) elements. If you are familiar with "lorem ipsum", the fake text generator that is used in software design as a placeholder for text, the idea is the same: generating placeholders for {shiny} outputs. For an example of an application built with {shinipsum}, please visit engineering-shiny.org/shinipsum/[5], or engineering-shiny.org/golemhtmltemplate/[6]: both these applications should look a little bit different every time you open them!

{shinipsum} can be installed from CRAN with:

---

[5]https://engineering-shiny.org/shinipsum/
[6]https://engineering-shiny.org/golemhtmltemplate/

```
install.packages("shinipsum")
```

You can install this package from GitHub with:

```
remotes::install_github("Thinkr-open/shinipsum")
```

This package includes a series of functions that generates random placeholders. For example, `random_ggplot()` generates random {ggplot2} (Wickham et al., 2021a) elements. If we run this code two times, we should get different results, as seen on Figure 9.2 and Figure 9.3.[7]

```
library(shinipsum)
library(ggplot2)
```

```
random_ggplot() +
  labs(title = "Random plot")
```

```
random_ggplot() +
  labs(title = "Random plot")
```

Of course, the idea is to combine this with a {shiny} interface, for example, `random_ggplot()` will be used with a `renderPlot()` and `plotOutput()`. And as we want to prototype but still be close to what the app might look like, these functions take arguments that can shape the output: for example, `random_ggplot()` has a `type` parameter that can help you select a specific {ggplot2} geom.

```
library(shiny)
library(shinipsum)
library(DT)
ui <- fluidPage(
```

---

[7]Well, there is a probability that we will get the same plot twice, and but that is the beauty of randomness.

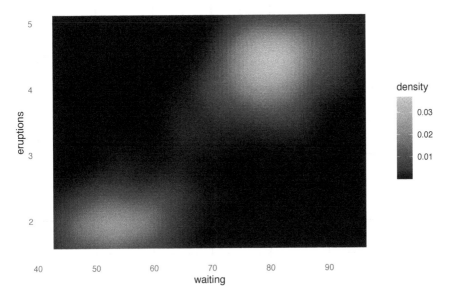

FIGURE 9.2: A random plot.

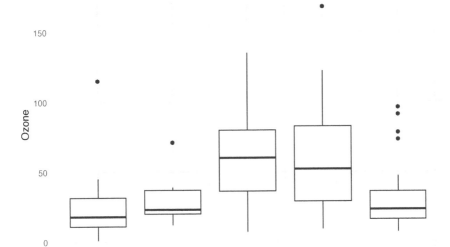

FIGURE 9.3: Another random plot.

```
  h2("A Random DT"),
  DTOutput("data_table"),
  h2("A Random Plot"),
  plotOutput("plot"),
  h2("A Random Text"),
  tableOutput("text")
)

server <- function(input, output, session) {
  output$data_table <- DT::renderDT({
    random_DT(5, 5)
  })
  output$plot <- renderPlot({
    random_ggplot()
  })
  output$text <- renderText({
    random_text(nwords = 50)
  })
}
shinyApp(ui, server)
```

Figure 9.4 is a screenshot of this application.

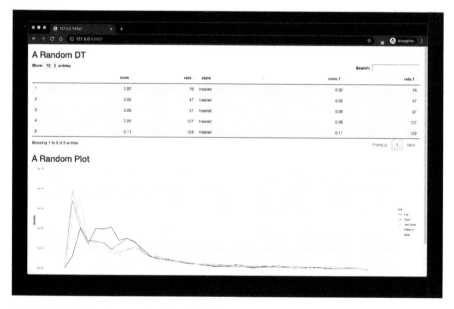

FIGURE 9.4: An app built with {shinipsum}.

Other {shinipsum} functions include:

- tables:

```
random_table(nrow = 3, ncol = 3)
```

```
        general picture blocks
general  24.641   5.991  33.52
picture   5.991   6.700  18.14
blocks   33.520  18.137 149.83
```

- print outputs:

```
random_print(type = "model")
```

```
Call:
lm(formula = Sepal.Length ~ Sepal.Width, data = datasets::iris)

Coefficients:
(Intercept)  Sepal.Width
      6.526       -0.223
```

and text, image, **ggplotly**, **dygraph**, and DT.

{shinipsum} is also a good tool if you want to demonstrate what a given UI framework will look like if used in {shiny}. This is, for example, what you find with {golemhtmltemplate}, available at engineering-shiny.org/golemhtmltemplate/[8], which uses a W3 web page template.[9]

9.2.2 Using {fakir} for fake data generation

Generating random placeholders for {shiny} might not be enough: maybe you also need example datasets.

This can be accomplished using the {fakir} package, which was primarily

---

[8]https://engineering-shiny.org/golemhtmltemplate/
[9]This application is also a demonstration of how to build a {golem} application using htmltemplate().

created to provide fake datasets for R tutorials and exercises, but that can easily be used inside a {shiny} application.

At the time of writing these lines, the package is only available on GitHub, and can be installed with:

```r
remotes::install_github("Thinkr-open/fakir")
```

This package contains three datasets that are randomly generated when you call the corresponding functions:

- `fake_base_clients()` generates a fake dataset for a ticketing service.
- `fake_sondage_answers()` is a fake survey about transportation.
- `fake_visits()` is a fake dataset for the visits on a website.

```r
library(fakir)
fake_visits(from = "2017-01-01", to = "2017-01-31")
```

```
# A tibble: 31 x 8
    timestamp  year month   day  home about  blog
  * <date>    <dbl> <dbl> <int> <int> <int> <int>
 1 2017-01-01  2017     1     1   369   220   404
 2 2017-01-02  2017     1     2   159   250   414
 3 2017-01-03  2017     1     3   436   170   498
 4 2017-01-04  2017     1     4    NA   258   526
 5 2017-01-05  2017     1     5   362    NA   407
 6 2017-01-06  2017     1     6   245   145   576
 7 2017-01-07  2017     1     7    NA    NA   484
 8 2017-01-08  2017     1     8   461   103   441
 9 2017-01-09  2017     1     9   337   113   673
10 2017-01-10  2017     1    10    NA   169   308
# ... with 21 more rows, and 1 more variable:
#   contact <int>
```

The idea with these datasets is to combine various formats that can reflect "real-life" datasets: they contain dates, numeric and character variables, and have missing values. They can also be manipulated with the included {sf} (Pebesma, 2021) geographical dataset `fra_sf` allowing for map creation.

Fake datasets created with {fakir} can be used to build light examples on the use of the inputs, for filters or interactive maps, or as examples for the internal functions and their corresponding documentation.

9.3 Building with RMarkdown

While on one side you are building the user interface, you (or someone from your team) can start working on the back-end implementation. This implementation should be done out of any reactive logic: the back-end should not depend on any reactive context. And because documentation is gold, you should start with writing the back-end documentation directly as package documentation:

- Inside your Vignettes folder: call `usethis::use_vignette()` to create the skeleton for a Vignette, which will then be used as package documentation.

- In the `inst/` folder, if you prefer not including these RMarkdown files as documentation for the end package.

Or what we call "Rmd-first".

9.3.1 Define the content of the application

Rmarkdown files are the perfect spot to sandbox the back-end of your application: inside the file, you don't have to think about any reactive behavior, as you are just working with plain old R code: data wrangling operations, multi-parameter-based models, summary tables outputs, graphical outputs, etc.

And the nice thing is that you can share the output of the rendered file as an HTML or PDF to either your client or boss, or anyone involved in the project. That way, **you can focus on the core algorithm**, not some UI implementation like "I want the button to be blue" when what you need to know is if the output of the model is correct. In other words, you are applying the rule of the separation of concerns, i.e. you help focus on one part of the application without adding any cognitive load to the person "reading" the outputs. And, last but not least, if you have to implement changes to the back-end functions, it is way easier to check and to share in a static file than in an application.

When doing that, the best way is again to separate things: do not be afraid of writing multiple RMarkdown files, one for each part of the end application. Again, this will help everybody focus on what matters: be it you, your team, or the person ordering the application.

Building the back-end in Rmd files is also a **good way to make the back-end "application independent"**, in the sense that it helps in documenting how the algorithms you have been building can be used outside of the application. In many cases, when you are building an application, you are creating functions that contain business logic/domain expertise, and that can, in fact, be used outside of the application. **Writing these functions and how they work together forces you to think about these functions, and also gives a good starting point for anybody familiar with R that would want to start using this back-end toolkit.** Of course, as you are building your application as a package, it is way easier now: you can share a package with the application inside it, along with a function to launch the app, but also functions that can be used outside.

And if you need some data to use as an example, feel free to pick one from {fakir}!

9.3.2   Using the Rmd files as a laboratory notebook

Rmd can also be used as the place to keep track of what you have in mind while creating the application: most of the time, you will create the functions inside the R/ folder, but it might not be the perfect place to document your thought process. On the other hand, using Markdown as a kind of "software laboratory notebook" to keep track of your idea is a good way to document all the choices you have made about your data wrangling, models, visualization, so that you can use it as a common knowledge-base throughout the application life: you can share this with your client, with the rest of your team, or with anybody involved in the project.

And also, developing in multiple Rmd files helps the separation of work between multiple developers, and will reduce code conflicts during development.

9.3.3   Rmd, Vignettes, and documentation first

Working with the {golem} (Fay et al., 2021) framework implies that you will build the application as an R package. And of course, an R package implies writing documentation: one of the main goals of the Vignettes, in an R package, is to document how to use the package. And the good news is that when checking a package, i.e. when running check() from {devtools} (Wickham et al., 2021d) or R CMD check, the Vignettes are going to be built, and the process will fail if at least one of the Vignettes fails to render. That way, you can use the documentation of the back-end as an extra tool for doing unit testing!

One radical approach to the "Rmd first" philosophy is to write **everything** in an Rmd from the very beginning of your project: write the function code, their roxygen tags, their tests, etc., then move everything to the correct spot in the package infrastructure once you are happy with everything. And of course, when you need to add another feature to your app, open a new markdown and start the process of development and documentation again.

Part IV

Step 3: Build

10

Building the App with {golem}

Building the App with {golem}

Now that the application is prototyped inside a {golem} (Fay et al., 2021) skeleton, you can work on its integration. In this step of the workflow, you will be linking the back-end and front-end together, and working on the global engineering of the application:

- add and organize dependencies
- create and include sub-modules if necessary
- organize utility functions and link them to the module in which they are used
- add testing infrastructure
- link to CI/CD services

Note that some concepts introduced here will be more extensively explored in the following chapters: the present chapter is a walkthrough of what you will find inside the 02_dev scripts.

10.1   Add dependencies

10.1.1   Package dependencies

When you are building a {shiny} (Chang et al., 2021a) application, you will have to deal with dependencies. Well, at least with one dependency, {shiny}. But chances are that you will not only be using {shiny} inside your application: you will probably call functions from other packages, for example, from {ggplot2} (Wickham et al., 2021a) for plotting, or any other package that is necessary for your application to work.

If you are building your application using {golem}, you will have 3 default dependencies: {golem} itself, {shiny}, and {config}. If you call golem::use_recommended_deps() in the first workflow script, you will also have {shiny}, {DT}, {attempt}, {glue}, {htmltools}, and {golem} as dependencies to your package.[1] But what about other dependencies like {ggplot2}? These ones need to be added by hand.

Here is how to process for a new dependency:

- Open the dev/02_dev.R script.
- Call   the   use_package()   function   from   {usethis}: usethis::use_package("pkg.you.want.to.add").
- Detail import mechanism in the related R files.

10.1.2   Importing packages and functions

There are two places where the dependencies of your application need to be managed:[2] the DESCRIPTION file and the NAMESPACE file.

- The DESCRIPTION file dictates which packages have to be installed **when your application is installed**. For example, when you install {golem} on your machine, you will also need other packages that are internally used by {golem}. And your application will also have dependencies: at the very least {shiny}, {golem}, and {config}. When building your application, you have to list these requirements somewhere, and the standard way to do that is by using the DESCRIPTION file.[3]

- The NAMESPACE file describes how your app interacts with the R session at run time, i.e. **when your application is launched**. With this NAMESPACE file, you can specify only a subset of functions to import from other packages: for example, you can choose to import only renderDT() and DTOutput() from {DT}, instead of importing all the functions. This selective import mechanism allows you to avoid namespace conflicts: for example, between jsonlite::flatten() and purrr::flatten().[4] To do so, we will need to go to every script that defines one or several function/s, and add a {roxygen2} (Wickham et al., 2020) tag, in the following form : #' @importFrom jsonlite fromJSON and #' @importFrom purrr flatten: that way, you are only importing fromJSON() from {jsonlite}.

Note that you can also use explicit namespacing, i.e. the pkg::function()

---

[1]The idea with this function is to provide a shortcut for adding commonly used dependencies, so that you don't have to do it by hand.

[2]This is not {shiny} or {golem} specific, but a requirement for any package.

[3]Note that most of the time, you will not be filling this by hand, but by using usethis::use_package().

[4]This can be pretty common as {jsonlite} might import JSON files as list, and {purrr} has pretty powerful tools for manipulating lists.

notation inside your code. And if you need a little help to identify dependencies, all the explicitly namespaced calls (pkg::function()) can be scraped using the {attachment} (Rochette and Guyader, 2021) package:

```
# This function will read all the scripts in the R/ folder and
# try to guess required dependencies
attachment::att_from_rscripts()
```

If you are using a development package (for example, one installed from GitHub), you can add it to the DESCRIPTION using the use_dev_package() function from {usethis}. This will add another field to the DESCRIPTION file, Remotes, with the location where the package is available.

All of this can seem a little bit daunting at first, but that is for the best:

Having a high quality namespace helps encapsulate your package and makes it self-contained. This ensures that other packages won't interfere with your code, that your code won't interfere with other packages, and that your package works regardless of the environment in which it's run.

R Packages (Wickham and Bryan, 2020)

To learn more about the details of how to manage dependencies, and about the DESCRIPTION and NAMESPACE files, here are some resources:

- *Writing R Extensions*[5], the official manual from the R-Core team
- R Packages[6], especially the *Package metadata* and *Namespace* chapters

10.2 Submodules and utility functions

When building a large application, you **will be splitting your codebase into smaller pieces.** In Chapter 3, "Structuring Your Project", that these

---

[5]https://cran.r-project.org/manuals.html
[6]https://r-pkgs.org/

utilitarian functions should be defined in files that are prefixed with a specific term. In the {golem} world, these are `utils_*` and `fct_*` files:

- `utils_*` files contain small functions that might be used several times in the application.
- `fct_*` files contain larger functions that are more central to the application.

Two functions can be called to create these files:

```
# Adding fct_ and utils_ files to the project
golem::add_fct( "helpers" )
golem::add_utils( "helpers" )
```

- The first will create a `R/fct_helpers.R` file.
- The second will create a `R/utils_helpers.R` file.

The idea, as explained before, is that as soon as you open a {golem}-based project, you are able to identify what the files contain, without having to open them.[7]

For example, the {hexmake} app has two of these files, `R/utils_ui.R`[8] and `R/utils_server.R`[9], in which you will find small functions that are reused throughout the app.

The `fct_*` files are to be used with larger functions, which are more central to the application, but that might not fit into a specific module. For example, in {hexmake}, you will find `R/fct_mongo.R`[10], which is used to handle all the things related to connecting and interacting with the Mongodb database.

As you can see, the difference is that `fct_*` file are more "topic centered", in the sense that they gather functions that relate to a specific feature of the application (here, the database), while `utils_*` files are more used as a place to put miscellaneous functions.

Note that when building a module with `golem::add_module()`, you can add a module-specific `fct_*` or `utils_*` file:

---

[7]The `utils_*` convention is a pretty common one: a lot of R packages contain a file called `utils.R` that bundles a series of small functions that are used throughout the package.

[8]https://github.com/ColinFay/hexmake/blob/master/R/utils_ui.R

[9]https://github.com/ColinFay/hexmake/blob/master/R/utils_server.R

[10]https://github.com/ColinFay/hexmake/blob/master/R/fct_mongo.R

```
# Creating the fct_ and utils_ file along the module creation
golem::add_module(
    name = "rendering",
    fct = "connect",
    utils = "wrapper"
)
```

Will create:

- R/mod_rendering.R
- R/mod_rendering_fct_connect.R
- R/mod_rendering_utils_wrapper.R

And this can also be done the other way around, by specifying the module you want to link your file to:

```
# Linking the utils_wrapper file to the rendering module
golem::add_utils("wrapper", module = "rendering")
```

10.3 Add tests

No piece of software should go into production if it has not been sufficiently tested. In this part of the building process, you will be setting tests for the application you are building. We will get back to the how, why and what of testing in an upcoming chapter, but as we are currently going through the 02_dev.R script, we mention here the line that allows you to add a test skeleton to your app.

If you have followed every step from the 01_start.R file, you already have a full testing infrastructure ready, with a set of recommended tests inserted by {golem}. But as it is hard to find tests that are relevant to all applications (as every application is unique), you will have to add and manually fill the tests that will check your app. And right now, to add a new testing file, you can call:

```
# Generate the testing infrastructure
usethis::use_test("app")
```

More on testing in Chapter 11.

10.4.1   Vignette

Vignettes are the long-format documentation for your application: users see this documentation when they are running `browseVignettes()`, when they look at the documentation in the `Help` pane from RStudio, or when they are browsing a web page on CRAN, and they are also the files that are used when the {pkgdown} websites are built. The good news is that if you have been using our "Rmd first" method, you already have most of the Vignettes built: they are the Markdown files describing how the back-end of your application works. Depending on how you applied this principle, these Rmd files might live inside the `inst/` folder, or already as package Vignettes. If you need to add a new Vignette, be it for adding an Rmd describing the back-end or a global documentation about the application, you can call the `use_vignette()` function from {usethis}.

```
# Adding a new Vignette named "shinyexample"
usethis::use_vignette("shinyexample")
```

Then, you can build all the Vignettes with:

```
# Compiling the Vignettes
devtools::build_vignettes()
```

10.4.2   Code coverage and continuous integration

A. Code coverage

Code coverage is a way to detect the volume of code that is covered by unit testing. You can do this locally, or you can use online services like Appveyor, an online platform that computes and tracks the code coverage of your repository.

To add it to your application, call the `use_coverage()` function from the {usethis} package:

```
# Adding the correct code coverage
# infrastructure in your application
usethis::use_coverage()
```

At the time of writing these lines, this function supports two services: Code-Cov[11] and coveralls[12].

Note that you can also perform code coverage locally, using the {covr} (Hester, 2020b) package, and the `package_coverage()` function.

```
# Compute the code coverage of your application
code_coverage <- covr::package_coverage()
```

For example, Figure 10.1 is the output of running the `package_coverage()` function on the {golem} package on the 2020-04-29 on the dev branch:

FIGURE 10.1: {golem} code coverage results.

As you can see, we reach a code coverage of almost 70%. **Deciding what the perfect percentage of coverage should be is not an easy task, and setting for an arbitrary coverage is not a smart move either, as it very much depends on the type of project you are working on.** For example, in {golem}, the addins.R file is not tested (0% code coverage), and that is for a good reason: these addins are linked to RStudio and are not

---

[11]https://codecov.io/
[12]https://coveralls.io/

meant to be tested/used in a non-interactive environment, and (at least at the time of writing these lines) there is no automated way to test for RStudio addins. Another thing to keep in mind while computing code coverage is that it counts the number of lines that are run when the tests are run, which means that if you write your whole function on one single line, you will have 100% code coverage. Another example is writing your `if/else` statement on one line `if (this) that else that`: your code coverage will count this line as covered, even if your test suite only runs the `if(this)` and not the `else`; in other words, even if your code coverage is good here, you are still not testing this algorithm extensively.

Note that you can also identify files with zero code coverage using the `covr::zero_coverage(covr::package_coverage())` function, which, instead of printing back a metric of coverage for each file, will point to all the lines that are not covered by tests inside your package, as shown in Figure 10.2.

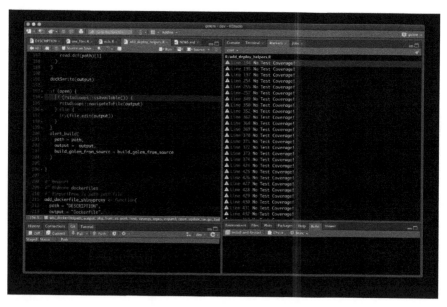

FIGURE 10.2: {golem} files with zero code coverage.

To sum up: do not set an arbitrary code coverage percentage goal, but rather use it as a general metric throughout your project. With CodeCov, you can get a timeline of the evolution of code coverage: a good tool for judging when you need to write more tests. For example, Figure 10.3 is the general tendency for the code coverage of the {tibble} package over the last 6 months (November 2019 to April 2020):

Perfect for getting a general feeling about the code coverage during the life of the project!

FIGURE 10.3: CodeCov.io results for the {tibble} package.

Note also that if you want to add the code coverage of your application inside a Vignette, you can use the {covrpage} (Sidi, 2021) package, which bundles the results of the {covr} coverage report into an interactive, human-readable Vignette that you can use later on as package documentation, or as an article inside your package website. {covrpage} can be installed from GitHub with `remotes::install_github('yonicd/covrpage')`.

B. Continuous Integration

Continuous integration, on the other hand, is ensuring the software is still working whenever a change is made by one of the developers. The idea is to add to the centralized version control system (for example, Git)[13] a service like Travis CI, GitHub Action (if you are on GitHub), or GitLab CI (for GitLab) that runs a series of commands whenever something is integrated in the repository, i.e. every time a change to the codebase is made. In other words, every time a new piece of code is sent to the central repository, a service runs regression tests that check that the software is still in a valid, working state.

You can set up various continuous integration services automatically by using functions from the {usethis} package:

---

[13]We will get back to version control in the Chapter 12, "Version Control"..

- Travis CI is set up with `usethis::use_travis()`.
- AppVeyor with `usethis::use_appveyor()`.
- GitLab CI with `use_gitlab_ci()`.
- Circle CI with `use_circleci()`.
- GitHub Actions with `use_github_actions()`.
- Jenkins with `use_jenkins()`.

If ever you want to add badges to your `README` files for these services, {usethis} also comes with a series of functions to do just that: `use_travis_badge()`, `use_appveyor_badge()`, `use_circleci_badge()` and `use_github_actions_badge()`.

CI services can do a lot more, like for example, deploy the application, build a container and send it to a container registry, compile RMarkdown files, etc.[14] The possibilities are almost limitless!

10.5 Using {golem} dev functions

When building an application, you will want it to behave differently depending on where it is run, and notably, if it is run in development or in production. We have seen in previous chapters that you can use the `golem-config.yml` file, or pass arguments to `run_app()`. A third option is to use the `dev` functions from {golem}.

There is a series of tools to make your app behave differently whether it is in "dev" or "prod" mode. Notably, the `app_prod()` and `app_dev()` functions look for the value of `options( "golem.app.prod" )`, or return `TRUE` if this option does not exist. In other words, by setting `options( "golem.app.prod" )` to `TRUE`, you will make the functions that depend on this option behave in a specific way.

Some functions pre-exist in {golem}, for example if you need to print a message to the console only during dev, you can do it with `cat_dev()`.

```
# Setting the option to FALSE
options( "golem.app.prod" = FALSE)
# Function runs as expected
golem::cat_dev("In dev\n")
```

---

[14]For example, the online version of this book is compiled to HTML every time something is merged into the `master` branch on GitHub.

In dev

```r
# Switching the option to TRUE
options( "golem.app.prod" = TRUE)
# Nothing is printing
golem::cat_dev("In dev\n")
```

Of course, chances are you do not only need to print things, you might want to use other functions. Good news! You can make any function being "dev-dependent" with the `make_dev()` function:

```r
# Same mechanism as cat_dev, but with other functions
log_dev <- golem::make_dev(log)
options( "golem.app.prod" = FALSE)
log_dev(10)
```

```
[1] 2.303
```

```r
options( "golem.app.prod" = TRUE)
log_dev(10)
```

That way, you can use functions in your back-end for development purposes, that will be ignored in production.

Part V

Step 4: Strengthen

11

Build Yourself a Safety Net

> Don't fuck over Future You
>
> *JD* (https://twitter.com/CMastication)

Strengthening your app means two things: testing and locking the application environment.

11.1 Testing your app

The process of getting your application production-ready implies that the application is tested. With a robust testing suite, you will develop, maintain, and improve in a safe environment and ensure your project sustainability. What will you be testing? Both sides of the application: the business logic and the user interface. And also, the application load, i.e. how much time and memory are required when your application starts being used by a significant number of users, be it from the user perspective (how long does it take to complete a full scenario) and from the server perspective (how much memory is needed for my app to run).

11.1.1 Testing the business logic

If you have been following the good practices we have listed in previous chapters, your current application has at least these two properties:

- The business-logic functions are separated from your interactive-logic functions.

- Your application is inside a package.

On top of being a sane organization approach, **using this separation inside a package structure allows you to leverage all the tooling that has been built for testing "standard" packages.**

R developers have been developing packages for a long time, and at the time of writing these lines (April 2020), more than 15,000 packages are available on CRAN. To sustain these developments, a lot of tools have been created to secure the development process, and especially tools for creating unit tests for your package.

Unit tests are a general concept in software engineering that describes the process of writing a form of assessment to check the validity of your code. A simplified explanation is that if you write a function called `meaning_of_life` that returns 42, you will expect this function to always return 42, and to be alerted if ever this value changes. Using unit tests is a way to secure your work in the future, be it for future you, for your collaborator, or for anybody wanting to collaborate on the project: if anyone comes and change the code behind the `meaning_of_life()` function, and the result is no longer 42, the developer working on this piece of code will be able to catch it. The general idea is to detect bugs and breaking changes at the moment they are happening, not once it is too late.

There are several packages in R that can be used to implement unit testing, and you can even implement your own tests. One of the most popular right now [1] is {testthat} (Wickham, 2021). This testing framework lets you write a series of tests and expectations, which are then launched when calling `test()` from {devtools} (Wickham et al., 2021d), either locally or in your CI system.

Here is an example of testing that the `meaning_of_life` will always be 42.

```
# Creating a testing context, with one expectation
test_that("The meaning of life is 42", {
  expect_equal(
    meaning_of_life(),
    42
  )
})
```

Once you have this test skeleton set, you will be able to detect any change to this function.

---

[1] Popularity based on the number of reverse dependencies and suggests, as shown in https://cran.r-project.org/web/packages/testthat/index.html.

If you want to learn more about how to use {testthat}, you can refer to the following resources:

- {testthat} online documentation[2]

- R Packages - Chapter 10 Testing[3]

- Building a package that lasts, eRum 2018 workshop - Part 5: Test and Code Coverage[4]

11.1.2   shiny::testServer()

At the time of writing these lines, the {shiny} team is actively working on a new way to test {shiny} server functions. These features are still a work in progress, and are not available in the stable version of {shiny} we have used in this book (1.6.0). Given that these features are still subject to change, we will not go into detail about these new features, but here is a preview of what it will look like:

```r
# Given the following module
computation_module_server <- function(input, output, session){
  ns <- session$ns
  r <- reactiveValues(
    value = NULL
  )
  observeEvent( input$selector , {
    r$value <- input$selector * 10
  })

}

# We can test it that way
library(shiny)
library(testthat)
testServer(computation_module_server, {

  # Give input$selector a value
  session$setInputs(selector = 1)
  # Call {testthat} functions
```

---

[2]https://testthat.r-lib.org/
[3]https://r-pkgs.org/tests.html
[4]https://speakerdeck.com/colinfay/building-a-package-that-lasts-erum-2018-workshop?slide=107

```
expect_equal(r$value, 10)

# Give input$selector a value
session$setInputs(selector = 2)
# Call {testthat} functions
expect_equal(r$value, 20)


})
```

This methodology is still under development, so we won't go deeper into this subject, but if you want to follow the update on this topic, please refer to the Server function testing with Shiny[5] article on the {shiny} website.

11.1.3   Testing the interactive logic

Once you have built a solid test suite for your business logic, another side of your app you might want to check is the interactive logic, i.e. the user interface.

There are several tools from the web development world that can be used to do exactly that: mimicking an interactive session where instead of deliberately clicking on the application interface, you let a program do it for you.

A. puppeteer

puppeteer is a NodeJS module that drives a Google Chrome headless session and mimics a session on the app.

And good news, there is a Google Chrome extension, called Puppeteer Recorder[6], that allows you to create, while visiting a web page, the pupepeteer script to reproduce your visit. Here is, for example, a very small JavaScript script for testing {hexmake} (Fay, 2021f), generated by this extension.

```
// Require the node module
const puppeteer = require('puppeteer');
(async () => {
```

[5]https://shiny.rstudio.com/articles/integration-testing.html
[6]https://chrome.google.com/webstore/detail/puppeteer-recorder/djeegiggegleadkkbgopoonhjimgehda

```
// launch puppeteer and connect to the page
const browser = await puppeteer.launch()
const page = await browser.newPage()
await page.goto('http://localhost:2811/')

// We're waiting for a DOM element to be ready
await page.waitForSelector('.row > .col > \
    .rounded > details:nth-child(3) > summary')
    // Now it's ready, we can click on it
    await page.click('.row > .col > .rounded > \
    details:nth-child(3) > summary')

    // Now our test is over, we can close the connection
    await browser.close()
})()
```

Be aware, though, that this extension does not record everything, at least with the version used while writing this book (0.7.1). For example, typing inside a text input is not recorded: that is completely doable inside **puppeteer**, yet not recorded by this extension.[7]

Once you have this piece of code, put it into a NodeJS script, and replay the session as many time as you need. If ever one of the steps cannot be replayed as recorded, the script will fail, notifying you of a regression.

Several packages in R mimic what **puppeteer** does (Google Chrome headless orchestration), with notably {crrri} (Lesur and Dervieux, 2021) and {chromote} (Chang, 2021). These packages can be used to launch and manipulate a Google Chrome headless session, meaning that you can programmatically navigate and interact with a web page from R. And to do the tests in a **puppeteer** spirit, you can refer to the {crrry} package (Fay, 2021b), which contains a series of wrapper functions around {crrri}, specifically designed for {shiny}.

Here is an example:

```
# Creating a new test instance
test <- crrry::CrrryOnPage$new(
  # Using the `find_chrome()` function to guess where the
  # Chrome bin is on our machine
  chrome_bin = pagedown::find_chrome(),
```

---

[7]See https://github.com/puppeteer/puppeteer/issues/441 for the code to set the text input values.

```
# Launching Chrome on a random available port on our machine
# Note that you will need httpuv >= 1.5.2 if you want to use
# this function
chrome_port = httpuv::randomPort(),
# Specifying the page we want to connect to
url = "https://connect.thinkr.fr/hexmake/",
# Do everything on the terminal, with no window open
headless = TRUE
)



Running \
  '/Applications/Google Chrome.app/Contents/MacOS/Google Chrome'
  --no-first-run --headless
  '--user-data-dir=/Users/colin/Library/Application Support/
    r-crrri/chrome-data-dir-dhutmfux'
  '--remote-debugging-port=40698'



# We'll wait for the application to be ready to accept inputs
test$wait_for_shiny_ready()



Shiny is computing
  Shiny is still running
```

You can then call one of the `test` object methods:

- `call_js()`, that allows you to run JavaScript code
- `shiny_set_input()` changes the value of a {shiny} Input
- `wait_for()` waits for a JavaScript condition to be TRUE
- `click_on_id` clicks on a given id

Of course, the interesting part is doing "bulk testing" of your application, for example, by setting a series of values to an input:

```
for (i in letters[1:5]){
  # We'll be setting a series of letters, one by one
  # for the package name input
  test$shiny_set_input(
    "main_ui_1-left_ui_1-pkg_name_ui_1-package",
    i
  )
}
```

```
-- Setting id main_ui_1-left_ui_1-pkg_name_ui_1-package--
Shiny is computing
  Shiny is still running
-- Setting id main_ui_1-left_ui_1-pkg_name_ui_1-package --
Shiny is computing
  Shiny is still running
-- Setting id main_ui_1-left_ui_1-pkg_name_ui_1-package --
Shiny is computing
  Shiny is still running
-- Setting id main_ui_1-left_ui_1-pkg_name_ui_1-package --
Shiny is computing
  Shiny is still running
-- Setting id main_ui_1-left_ui_1-pkg_name_ui_1-package --
Shiny is computing
  Shiny is still running
```

And once your test is done, do not forget to close the connection!

```
# Closing the connection
test$stop()
```

B. Monkey test

If you are working on a user-facing software (i.e. a software used by external users), there is one rule to live by: every unexpected behavior that can happen, will happen. In other words, if you develop and think "a user will never do that", just expect a user to eventually do "that".

But how can we get prepared for the unexpected? How can we test the "crazy behavior" that user will adopt? In web development, there exists a methodology called "Monkey testing", which consists of **launching a series of random event on a web page: random text in input, scrolling, clicking, zooming... and see if the application crashes or not**. This software testing method allows you to test the robustness of the application, by seeing how well it can handle unexpected behaviors.

Several JavaScript libraries exist when it comes to monkey testing, one of the most popular (and easy to use) libraries is called `gremlin.js`[8]. This library is particularly interesting when it comes to {shiny} as it does not need external installation: you can add the library as a bookmark on your browser, navigate to the application, and launch the testing (click on the "Generate

---

[8]https://github.com/marmelab/gremlins.js

Bookmarklet" link on the top of the README[9]). Figure 11.1 show an example of running gremlins on the prenoms application.

FIGURE 11.1: Example of using `gremlins.js` on the "prenoms" {shiny} application.

And if you want to scale this, you can also combine it with {shinyloadtest} (Schloerke et al., 2021): launch a session recording, run **gremlins** one or several time inside the recording, then replay it with multiple sessions.

With {crrry}, this **gremlins** test comes for free:

```
# Creating a new test instance
test <- crrry::CrrryOnPage$new(
    # Using the `find_chrome()` function to guess where the
    # Chrome bin is on our machine
    chrome_bin = pagedown::find_chrome(),
    # Launching Chrome on a random available port on our machine
    # Note that you will need httpuv >= 1.5.2 if you want to use
    # this function
    chrome_port = httpuv::randomPort(),
    # Specifying the page we want to connect to
    url = "https://connect.thinkr.fr/hexmake/",
    # Do everything on the terminal, with no window open
```

[9](https://github.com/marmelab/gremlins.js)

```
  headless = TRUE
)
# We'll wait for the application to be ready to accept inputs
test$wait_for_shiny_ready()
# We launch the horde of gremlins
test$gremlins_horde()
# Sleep, let the gremlins do their job
Sys.sleep(10)
# Check that the app is still working
test$wait_for_shiny_ready()
# Stop the connection
test$stop()
```

C. {shinytest}

Finally, if you prefer a {shiny} specific package, you can go for {shinytest} (Chang et al., 2021b). This package, created and maintained by RStudio, **allows you to do a series of screenshots of your application, and then replays your app and compares the previously taken screenshots to the current state of your application**, allowing you to detect any changes in the interface.

If you are building your application with {golem} (Fay et al., 2021), you will need to add an **app.R** file at the root of your package, then run **shinytest::recordTest()**:

```
# Create an app.R file at the root of the package
golem::add_rstudioconnect_file()
# Launch a test, and record a series of
# snapshots of your application
shinytest::recordTest()
```

Once this function is run, a new window opens: it contains your app, and a "Screenshot" button on the right. Using this button, you can take various recordings of your shiny application at different states, as shown in Figure 11.2.

Then, you can do some changes in your app, and run:

```
shinytest::testApp()
```

FIGURE 11.2: General view of a {shinytest} window.

If the {shinytest} package detects a visual change in the application, you
will be immediately alerted, with a report of the difference from the snapshots
you took and the current state of the application.

11.1.4   Testing the app load

A. {shinyloadtest}

{shinyloadtest} (Schloerke et al., 2021) **tests how an application behaves when one, two, three, twenty, one hundred users connect to
the app and use it,** and gives you a visual report about the connection and
response time of each session. The idea with {shinyloadtest} is to first record
a session where you mimic a user behavior, then **shinycannon**, a command line
tool that comes with {shinyloadtest}, replays the recording several times.
Once the session has been replayed several times mimicking the session you
have recorded, you have access to a report of the behavior of your app.

```
library(shinyloadtest)
```

```r
# Starting your app in another process
p <- processx::process$new(
  "Rscript",
  c( "-e", "options('shiny.port'= 2811);hexmake::run_app()" )
)
# We wait for the app to be ready
Sys.sleep(5)
# Check that the process is alive
p$is_alive()
# Open the app in our browser just to be sure
browseURL("http://localhost:2811")
```

Record the tests, potentially in a new dir:

```r
# Creating a directory to receive the logs
fs::dir_create("shinylogs")
# Performing the session recording inside this new folder
withr::with_dir(
  "shinylogs", {
    # Launch the recording of an app session, using port 1234
    shinyloadtest::record_session(
      "http://localhost:2811",
      port = 1234
    )
  }
)
```

We now have a series of one or more recording/s inside the **shinylogs/** folder:

Then, let's switch to our command line, and rerun the session with **shinycannon**. The **shinycannon** command line tools take several arguments: the path the **.log** file, the URL of the app, **--workers** specify the number of concurrent connections to run, and the **--output-dir** argument specifies where the report should be written.

Then, go to your terminal and run:

```
shinycannon shinylogs/recording.log \
  http://localhost:2811 --workers 10 \
  --output-dir shinylogs/run1
```

And now, we have new files inside the folder, corresponding to the session recordings.

```
# printing the structure of shinylogs
fs::dir_tree("shinylogs", recurse = FALSE)
```

```
shinylogs
+-- recording.log
+-- run1
+-- run2
\-- run3
```

Good news: we do not have to manually analyze these files—{shinyloadtest} offers a series of wrapper functions to do that.

```
# Bringing the runs in the R session
shinyload_runs <- shinyloadtest::load_runs(
  "5 workers" = "shinylogs/run1"
)
```

We now have a data.frame that looks like this:

```
dplyr::glimpse(head(shinyload_runs))
```

```
Rows: 6
Columns: 13
$ run               <ord> 5 workers, 5 workers, 5 wor~
$ user_id           <int> 0, 0, 0, 0, 0, 0
$ session_id        <int> 0, 0, 0, 0, 0, 0
$ iteration         <int> 0, 0, 0, 0, 0, 0
$ input_line_number <int> 4, 5, 6, 7, 8, 9
$ event             <chr> "REQ_HOME", "WS_OPEN", "WS_~
$ start             <dbl> 0.000, 0.462, 0.539, 1.024,~
$ end               <dbl> 0.461, 0.539, 0.542, 1.024,~
$ time              <dbl> 0.461, 0.077, 0.003, 0.000,~
$ concurrency       <dbl> 0, 1, 1, 1, 1, 1
$ maintenance       <lgl> TRUE, TRUE, TRUE, TRUE, TRU~
$ label             <ord> "Event 1) Get: Homepage", "~
$ json              <list> ["REQ_HOME", 2020-04-10 12:~
```

Then, {shinyloadtest} comes with a series of plotting functions that can be
used to analyze your recording. For example:

- `slt_session_duration()` plots the session duration, with the various
  types of events that take computation time: JS and CSS load, R computation, etc. The output is available in Figure 11.3.

```
slt_session_duration(shinyload_runs)
```

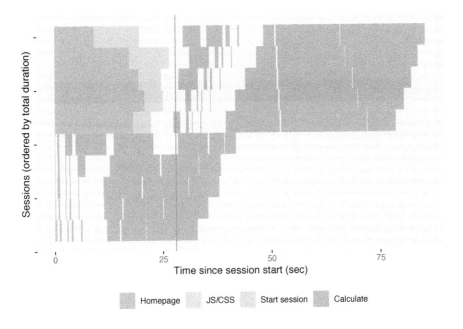

FIGURE 11.3: Session duration.

And if you need to bundle everything into an HTML reports,
`shinyloadtest_report()` is what you are looking for.

```
# Generating the report
shinyloadtest_report(shinyload_runs)
```

This function will generate an HTML report of all the things computed by
{shinyloadtest}, as shown in Figure 11.4.

To sum up with a step-by-step guide:

FIGURE 11.4: Webpage generated by `shinyloadtest_report()`.

- If the shiny app is only available locally, on your machine, then launch a process with {processx} (Csárdi and Chang, 2021), or in another R session, that launches the application. You can either set the port with `options('shiny.port'= 2811)`, or let shiny decide for you. Be sure that the process is running. If the app is online, use the online URL (and make sure you have access to the app).

- Run `shinyloadtest::record_session(url)`. You should probably set a different port for {shinyloadtest}, so that it does not try to connect on port 80.

- Play around with your app; record a scenario of usage.

- Close the tab where the app is running.

- Return to your terminal, and run the **shinycannon** command line tool.

- Wait for the process to be finished.

- Go back to R, and then you can analyze the data from the recordings, either manually or by generating the HTML report.

B. {shinyloadtest}, {crrry}, and {dockerstats}

Another thing you might want to monitor is the memory/CPU usage of your application, which {shinyloadtest} does not natively provide: the package records the load from the browser point of view, not from the server one. That's where {dockerstats} (Fay, 2021c) can come into play: this package is a wrapper around the command line docker stats, and returns an R data.frame with the stats.

You can get the {dockerstats} package from GitHub with:

```
remotes::install_github("ColinFay/dockerstats")
```

Or from NPM via:

```
npm install -g r-dockerstats
```

```
library(dockerstats)
```

With these stats, we can monitor the load on the app when it is run in a Docker container.

We will start by launching the container using a system() call: here, we are running the {hexmake} application, bundled in the colinfay/hexmake Docker image, on port 2811. We also make sure we give it a name with --name, so that we can call it in our dockerstats() call later on.

```
# We are launching the docker container
# using R system() command. Here, we are
# running the container image called
# colinfay/hexmake. We are naming the
# container hexmake using the --name flag,
# --rm means the container will be removed
# when stopped, and finally the -p flag defines
# how to bind the ports of the container
# with the ports of the host (left is the host,
#  right is the container): in other word, here,
#  we bind port 80 of our container to the port 2811
# of our machine.
system(
  "docker run --name hexmake --rm -p 2811:80 colinfay/hexmake",
  wait = FALSE
)
```

Let's say now we want the stats for the hexmake container:

```r
# Waiting for the container to be ready
Sys.sleep(30)
# Showing the docker stats for hexmake
tibble::as_tibble(
  dockerstats("hexmake")
)
```

```
# A tibble: 1 x 14
  Container Name      ID        CPUPerc MemUsage MemLimit
  <chr>     <chr>     <chr>       <dbl> <chr>    <chr>
1 hexmake   hexmake   34cc1544~    0.37 108.5MiB 5.807GiB
# ... with 8 more variables: MemPerc <dbl>,
#   NetI <chr>, NetO <chr>, BlockI <chr>,
#   BlockO <chr>, PIDs <int>, record_time <chr>,
#   extra <chr>
```

Of course, right now nobody is using the app, so the usage can be pretty small. But let's push it a little bit by mimicking a lot of connections.

To do that, we can replay our **shinycannon** call, and at the same time use the `dockerstats_recurse()` function, which will recursively call `dockerstats()` on a regular interval.

```r
# Replaying the recording
shinycannon shinylogs/recording.log \
# Specificying the host url and the number of "visitors"
  http://localhost:2811 --workers 10 \
# Define where the recording will be outputed
  --output-dir shinylogs/run3
```

Let's launch at the same time a `dockerstats_recurse()`. For example, here, we will print, on each loop, the `MemUsage` of the container, then save the data inside a `dockerstats.csv` file.

```r
# Calling recursive the dockerstats function.
# The callback function takes a function, and define
# what to do with the data.frame each time the
# dockerstats results are computed.
```

```
dockerstats_recurse(
  "hexmake",
  # append_csv is a {dockerstats} function that will
  # apped the output to a given csv
  callback = append_csv(
    file = "shinylogs/dockerstats.csv",
    print = TRUE
  )
)
```

Figure 11.5 shows these processes side to side.

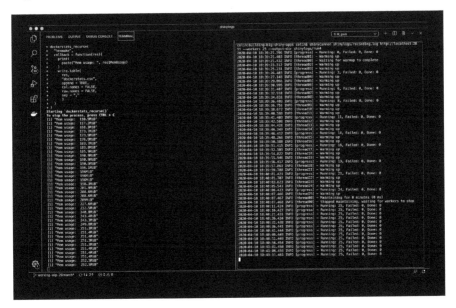

FIGURE 11.5: {dockerstats} and `shinycannon` running side-by-side at the same time.

As you can see, as the number of connections grow, the memory usage grows. And we now have a csv with the evolution of the `docker stats` records over time!

```
# read_appended_csv() allows to read a csv that has been
# constructed with the append_csv() function
docker_stats <- read_appended_csv(
  "shinylogs/dockerstats.csv"
)
```

```
dplyr::glimpse(head(docker_stats))
```

```
Rows: 6
Columns: 14
$ Container   <chr> "hexmake", "hexmake", "hexmake", ~
$ Name        <chr> "hexmake", "hexmake", "hexmake", ~
$ ID          <chr> "b5d337941e310cbf4708b95a9cc75560~
$ CPUPerc     <dbl> 0.09, 15.07, 38.58, 54.94, 20.28,~
$ MemUsage    <chr> "110.9MiB", "117.2MiB", "168.3MiB~
$ MemLimit    <chr> "1.943GiB", "1.943GiB", "1.943GiB~
$ MemPerc     <dbl> 5.57, 5.89, 8.46, 8.73, 8.74, 8.89
$ NetI        <chr> "586B", "8.37kB", "31.6kB", "62.1~
$ NetO        <chr> "0B", "433kB", "1.18MB", "2.48MB"~
$ BlockI      <chr> "0B", "0B", "0B", "0B", "0B", "0B"
$ BlockO      <chr> "0B", "0B", "0B", "0B", "0B", "0B"
$ PIDs        <int> 3, 4, 4, 4, 4, 4
$ record_time <dttm> 2020-04-10 10:39:20, 2020-04-10 1~
$ extra       <lgl> NA, NA, NA, NA, NA, NA
```

If you need a deeper look into the connection between application actions and the Docker stats, you can also combine {dockerstats} with {crrry}, the idea being that you can record the CPU usage at the exact moment the application performs a specific computation.

Let's record the computation of the **hexmake** container containing the same app as before.

First, launch the container:

```
# Launching the container a second time,
# but name it xmk2 and serve it on port 2708
system(
  "docker run -p 2708:80 --rm --name xmk2 -d colinfay/hexmake",
  wait = FALSE
)
Sys.sleep(60) # Let the container launch
```

Then, a {crrry} job:

```
# See previous version of this code for a commented explanation
test <- crrry::CrrryOnPage$new(
  chrome_bin = pagedown::find_chrome(),
  chrome_port = httpuv::randomPort(),
  url ="http://localhost:2708",
  headless = TRUE
)
```

```
Running '/Applications/Google
  Chrome.app/Contents/MacOS/Google Chrome'
  --no-first-run --headless \
  '--user-data-dir=/Users/colin/Library/Application
  Support/r-crrri/chrome-data-dir-thyhpptv' \
  '--remote-debugging-port=48938'
```

```
test$wait_for_shiny_ready()
```

```
Shiny is computing
  Shiny is still running
```

```
# We are creating a first data.frame that records the launch
# of the container.
results <- dockerstats::dockerstats("xmk2", extra = "launch")

for (i in letters[1:10]){
  # We will be setting a letter for the package name input
  test$shiny_set_input(
    "main_ui_1-left_ui_1-pkg_name_ui_1-package",
    i
  )
  # Once the input is set, we call dockerstats()
  # for this container and bind the results to
  # the previously created data.frame
  results <- rbind(
    results,
    dockerstats::dockerstats("xmk2", extra = i)
  )
}
```

```
-- Setting id main_ui_1-left_ui_1-pkg_name_ui_1-package --
Shiny is computing
  Shiny is still running
-- Setting id main_ui_1-left_ui_1-pkg_name_ui_1-package --
Shiny is computing
  Shiny is still running
-- Setting id main_ui_1-left_ui_1-pkg_name_ui_1-package --
Shiny is computing
  Shiny is still running
-- Setting id main_ui_1-left_ui_1-pkg_name_ui_1-package --
Shiny is computing
  Shiny is still running
-- Setting id main_ui_1-left_ui_1-pkg_name_ui_1-package --
Shiny is computing
  Shiny is still running
-- Setting id main_ui_1-left_ui_1-pkg_name_ui_1-package --
Shiny is computing
  Shiny is still running
-- Setting id main_ui_1-left_ui_1-pkg_name_ui_1-package --
Shiny is computing
  Shiny is still running
-- Setting id main_ui_1-left_ui_1-pkg_name_ui_1-package --
Shiny is computing
  Shiny is still running
-- Setting id main_ui_1-left_ui_1-pkg_name_ui_1-package --
Shiny is computing
  Shiny is still running
-- Setting id main_ui_1-left_ui_1-pkg_name_ui_1-package --
Shiny is computing
  Shiny is still running
```

And draw a small graph of this evolution, shown in Figure 11.6:

```
library(dplyr, warn.conflicts = FALSE)
# We are converting the MemUsage and record_time columns
# to a format that can be used in {ggplot2}
results <- results %>%
  mutate(
    MemUsage = to_mib(MemUsage),
    record_time = as.POSIXct(record_time)
  )
library(ggplot2)
# Using the record time as an x axis,
# then adding the MemUsage as a line (to watch the
```

```
# evolution over time), then we add the 'extra' column,
# that contains the letters, as vertical lines + as
# label
ggplot(
  data = results,
  aes(x = record_time)
) +
  geom_line(
    aes(y = MemUsage)
  ) +
  geom_vline(
    aes(xintercept = record_time)
  ) +
  geom_label(
    aes(
      y = max(MemUsage),
      label = extra
    )
  ) +
  labs(
    title = "MemUsage of 10 inputs for package name"
  )
```

11.2 A reproducible environment

One of the challenges of building an app that needs to be sent to production is that you will need to work in a reproducible environment. What does this mean? **When building a production application, you are building a piece of software that will be launched on another computer, be it a server in the cloud or someone else's computer**. Once your app is built, there are few chances that you will launch it on your own computer and that external users will connect to your computer. You will either give your users a package (which will be the simplest way to share it: bundle the packaged app to a `tar.gz`, then let people install it either manually or from a package repository), or a URL where they can connect and use your app.

If you follow the {golem} workflow and all the good practices for a solid package, the application you have built should be deployable on another computer that has R. In that second case, **you will have to think about how you can create your app in a reproducible environment**: in other words,

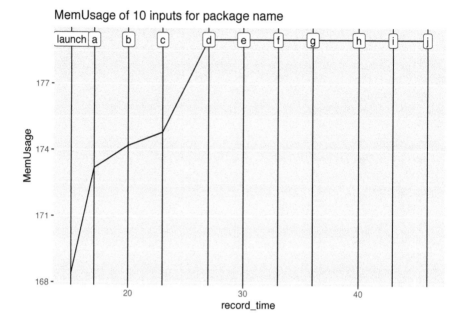

FIGURE 11.6: Plot of the {dockerstats} evolution.

be sure that the app is deployed under the same configuration as your local application—R version, package versions, system requirements, environment variables, etc.

To help you achieve that, we will introduce two tools in the next section: {renv} (Ushey, 2020) and Docker[10].

11.2.1 {renv}

A. About {renv}

How do we make sure the package versions we have installed on our machine stay the same in the production environment? And also, how can we be sure that, working as a team, we will be able to work together using the same package versions?

From one package version to another, functions and behaviors change. Most of the time, a new version means new functions and new features. But from time to time, a new version means breaking changes. **Monitoring these changes and how they potentially break our code is a hard task: because**

---

[10]https://www.docker.com/

checking versions of packages on various machines can take time, or
because debugging these bugs in your application logs is not straightforward.
And of course, the moment when we discover the error might not be the
perfect time for us, as we might not have enough free time on our calendar to
debug the application that has stopped running.

Let's take, for example, this traceback from the logs of an application we sent
one day on a `Shiny Server`:

```
root@westeros-vm:/var/log/shiny-server# cat thewall(...).log
*** caught segfault ***
[...]
address 0x5100004d, cause 'memory not mapped'

Traceback:
1: rcpp_sf_to_geojson(sf, digits, factors_as_string)
2: sf_geojson.sf(data)
3: geojsonsf::sf_geojson(data)
4: addGlifyPolygons(., data = pol_V1, color = les_couleurs,
popup = "val", opacity = 1)
5: function_list[[i]](value)
6: freduce(value, `_function_list`)
7: `_fseq`(`_lhs`)
8: eval(quote(`_fseq`(`_lhs`)), env, env)
[...]
105: captureStackTraces({
while (!.globals$stopped) {
..stacktracefloor..(serviceApp())
Sys.sleep(0.001)    }})
106: ..stacktraceoff..(captureStackTraces({
while (!.globals$stopped) {
..stacktracefloor..(serviceApp())
Sys.sleep(0.001)    }}))
107: runApp(Sys.getenv("SHINY_APP"),
port = port,
launch.browser = FALSE)
An irrecoverable exception occurred. R is aborting now ...
```

Pretty hard to debug, isn't it? What has actually happened? On that specific
case, it turned out that the package version from {geojsonsf} (Cooley, 2020)
was `1.2.1` on our development machine, and the one on the {shiny} server
was updated to `1.3.0`, and there was a breaking change in the package, as
shown in Figure 11.7. This bug was hard to detect as {geojsonsf} was not a

direct dependency of our app, but a dependency of one of our dependencies, making it slightly more complex to identify.

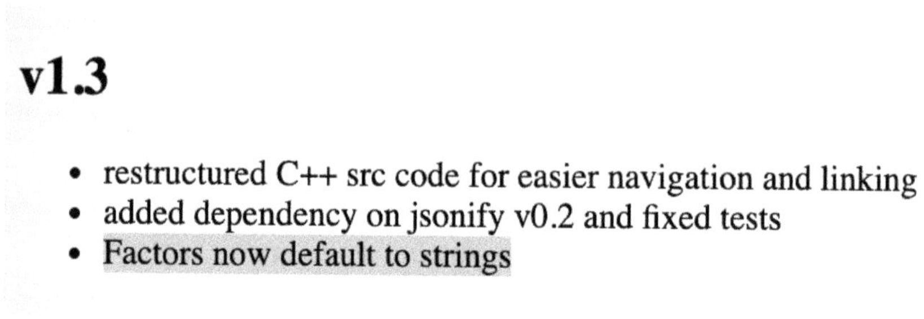

v1.3

- restructured C++ src code for easier navigation and linking
- added dependency on jsonify v0.2 and fixed tests
- Factors now default to strings

FIGURE 11.7: Breaking changes in {geojsonsf}, a dependency of a dependency of our {shiny} application.

The same thing could have happened if working as a team: one of the computers has an old version, when another one has updated to a more recent one. How do we prevent that? This is where the {renv} package comes into play: this package allows you to have a project-based library, instead of a global one. In other words, instead of having a library that is global to your machine, {renv} allows you to specify packages with fixed versions for a project. That means that you can have {geojsonsf} version 1.2.1 in one of your projects, and version 1.3.0 in another, with the two not conflicting with each other.

B. Using {renv}

Underlying the philosophy of renv is that any of your existing workflows should just work as they did before.

Introduction to renv (https://rstudio.github.io/renv/articles/renv.html)

The first thing to do with {renv} is initiate it with the init() function.

```
# Loading and initiaing {renv}
library(renv)
init()
```

This function does several things:

- Create/modify the .Rprofile file at the root of your project. Here is an example of what this file may look like inside an empty project:

```
source("renv/activate.R")
```

In this example, there is just one call to a script, one located at renv/activate.R.

- It creates a renv.lock file, which will list all the package dependencies

As we have initiated an empty project, we do not have any dependencies here. If you run this command in a project that already has scripts and dependencies, {renv} will try to locate them all, and add them to this file. Note that these packages may come from CRAN, Bioconductor, GitHub, GitLab, Bitbucket, and even local repositories.

The renv/ folder contains a series of files that store your settings and the necessary packages, using a structure that mimics a local repository.

```
# Displaying the structure of the folder
fs::dir_tree("data-raw/renvinit/", recurse = 5)
```

```
data-raw/renvinit/
+-- renv
|   +-- activate.R
|   +-- library
|   |   \-- R-3.6
|   |       \-- x86_64-apple-darwin15.6.0
|   |           \-- renv
|   |               +-- DESCRIPTION
|   |               +-- INDEX
|   |               +-- LICENSE
|   |               +-- Meta
|   |               +-- NAMESPACE
|   |               +-- NEWS.md
|   |               +-- R
|   |               +-- doc
|   |               +-- help
|   |               +-- html
|   |               \-- resources
|   \-- settings.dcf
+-- renv.lock
\-- renvinit.Rproj
```

We will not go into details on this folder, as it is a rather complex structure and chances are that you will never have to update it by hand.

With {renv}, you can choose to link this "local repository" to a local cache, i.e. a folder which is common to all your projects and stores packages and the different versions you already installed (this is the default behavior) or to store the complete packages inside the project, making it portable.

When you need a new package, you will have to install it in your local library. The fastest way to install new packages in your {renv}-powered project is by using the `install.packages` function, which is shimmed by {renv}. This shim will search the local cache to see if the package has already been cached, and if it is not, it will install and link it.

Now, we need to install a new package, for example {attempt} (Fay, 2020):

```r
# Installing attempt
install.packages("attempt")
```

We will now add a little call to this library:

```r
# Create a fake script that launches {attempt}
write("library(attempt)", "script.R")
```

Once you want to update your {renv} `Lockfile`, call `snapshot()`.

```r
# Snapshoting the current status of the environment
renv::snapshot(confirm = FALSE)
```

Note that if you are building an application as a package, use `renv::snapshot(type = "explicit")` (need version > 0.9.3-99): this will only capture the dependencies listed in the `DESCRIPTION` file. If you don't specify this `type = "explicit"`, {renv} will go and look for all the packages it can find in your file, and notably in your `.Rhistory`, meaning that if you one day used a function from {dplyr} but your current package doesn't use it anymore, {renv} will include it.

```json
{
  "R": {
    "Version": "3.6.1",
    "Repositories": [
```

```
          {
            "Name": "CRAN",
            "URL": "https://cran.rstudio.com"
          }
       ]
   },
   "Packages": {
     "attempt": {
       "Package": "attempt",
       "Version": "0.3.0",
       "Source": "Repository",
       "Repository": "CRAN",
       "Hash": "9aaae25e273927dba4e279caac478baa"
     },
     "renv": {
       "Package": "renv",
       "Version": "0.9.3",
       "Source": "Repository",
       "Repository": "CRAN",
       "Hash": "c1a367437d8a8a44bec4b9d4974cb20c"
     },
     "rlang": {
       "Package": "rlang",
       "Version": "0.4.5",
       "Source": "Repository",
       "Repository": "CRAN",
       "Hash": "1cc1b38e4db40ea6eb19ab8080bbed3b"
     }
   }
}
```

And now that you have a reproducible {renv} library, what is next? Of course, if you are either working as a team or deploying to a server, you will have to restore the state of your project, which is now living somewhere else, inside your current project/deployment. And to do that, the function to call is renv::restore(), which will update your local project with the dependencies listed inside your Lockfile.

To sum up, here are the steps to follow:

- Initiate the project with renv::init().
- Install/remove packages.
- Take a snapshot() of the state of your project.
- renv::restore() the state of your project using renv.lock.
- Share .Rprofile, renv.lock, renv/activate.R and renv/settings.dcf files for reproducibility.

Of course, `renv::restore()` comes with another superpower: time traveling! If you decide to update a package in your project, and realize that this package makes the application crash (e.g., an update to {geojsonsf}), you can go back in time to a previous version of your library by calling the `restore()` function.

There are more things you can do with {renv}. If you want to know more, we invite you to refer to the official documentation[11].

11.2.2   Docker

A. R, Docker, {shiny}

Docker is a program that allows to download, install, create, launch and stop multiple operating systems, called containers, on a machine, which will be called the host. This host can be your local computer, or the server where you deploy your application/s.

Docker was designed for **enclosing software environments inside an image that can later be launched**. The general idea is that with Docker, you are defining in a `Dockerfile` all the "rules" that are used to create a given environment, and then you can use this file (and the linked files, for example the R package containing your app) to **deploy your application on any given server that can run Docker**. That way, if the `Dockerfile` can compile on your machine and if you can run it, it should work everywhere (of course, it is a little bit more complex than that, but you get the idea).

Why Docker in the context of {shiny} apps? Because Docker allows you to abstract away the complexity of managing multiple versions of R and multiple versions of the same package, or even different versions of the same system requirement. For example, let's take our example with the breaking change in {geojsonsf} that we used in the previous section. With Docker, we can safely specify a `1.2.1` version in our image, and changing versions on the server would not have broken our code.

By using Docker for your deployment, you can build and deploy an application with the very same version of packages and R as the one on your computer. And of course, you can change them without breaking the rest of the machine: everything that happens in a container stays in a container. That way, if you are building your application with an older version of {shiny}, **you are sure that sending it to production will not break everything: the version inside the Docker image is the same as the one from your machine.** And later, if you update {shiny} and start a new project, you can deploy your app with another version of the package. Same goes for your version of R.

---

[11]https://rstudio.github.io/renv

B. Building a Dockerfile for your app

Good news! If you are building your app with {golem}, the creation of the Dockerfile is just one function away! If you have a look at the 03_deploy.R file in the dev folder, you will find a series of functions that can create the Dockerfile for your project: either as a generic Docker image, or for {shiny}Proxy or Heroku.

For example, to create a Dockerfile for a {golem} project, you can run the following, from the root of your package:

```
golem::add_dockerfile()
```

Let's take some time to understand this file, and detail how we could be building it from scratch.

1. FROM

```
FROM rocker/r-ver:4.1.0
```

This line defines what version of R to use for deploying your application. This FROM line is the one that sets an image to start from: you rarely (if ever) build a Docker image from nothing, but instead you use an existing image on top of which you build your own image. Here, we choose one of the r-ver[12] Docker images, based on the output of:

```
R.Version()$version.string
```

```
[1] "R version 4.1.0 (2021-05-18)"
```

2. RUN

The RUN call in the file refers to bash calls that are used to build the image. For example, the second line of the Dockerfile installs all the system requirements needed by our application.

```
RUN apt-get update && \
  apt-get install -y  git-core \
  libcurl4-openssl-dev libssh2-1-dev \
  libssl-dev libxml2-dev make \
  zlib1g-dev && rm -rf /var/lib/apt/lists/*
```

---

[12]https://hub.docker.com/r/rocker/r-ver/

In the subsequent RUN calls, {golem} chooses to call
remotes::install_version() to be sure we install the version of the
package that matches the one from your computer.

```
RUN Rscript -e \
'remotes::install_version("xfun",upgrade="never",version="0.19")'
```

As you can see, it matches the local version:

```
packageVersion("xfun")
```

```
[1] '0.19'
```

3. ADD

This Docker entry takes a folder or a file, and copies it inside the image. With
{golem}, we are adding the current project, containing the app, to a folder
called /build_zone.

```
ADD . /build_zone
```

4. EXPOSE

This command defines which port of the container will be available from the
outside of the container.

```
EXPOSE 80
```

5. CMD

This final command is the one that is launched when you run a container.
With a {shiny} app, this command is the one that launches the application.

```
CMD R -e \
  "options('shiny.port'=80,shiny.host='0.0.0.0');golex::run_app()"
```

C. {dockerfiler}

If you want to do everything from the R command line, the {dockerfiler} (Fay, 2019) package is here for you! This package allows you to generate a Dockerfile straight from R:

```r
library(dockerfiler)
# Creating a new Dockerfile object
my_dock <- Dockerfile$new()
# Adding RUN, ADD, WORKDIR and EXPOSE commands
my_dock$RUN("apt-get update && apt-get install -y git-core")
my_dock$ADD(".", "/")
my_dock$RUN("mkdir /build_zone")
my_dock$ADD(".", "/build_zone")
my_dock$WORKDIR("/build_zone")
my_dock$RUN(r(remotes::install_local(upgrade="never")))
my_dock$EXPOSE(80)
# Viewing the Dockerfile
my_dock
```

```
FROM rocker/r-base
RUN apt-get update && apt-get install -y git-core
ADD . /
RUN mkdir /build_zone
ADD . /build_zone
WORKDIR /build_zone
RUN R -e 'remotes::install_local(upgrade = "never")'
EXPOSE 80
```

D. Docker and {renv}

If you use {renv} to build your {shiny} application, it can also be used inside your Docker container. To make those two tools work together, you will have to copy the files produced by {renv} inside the container: .Rprofile, renv.lock, renv/activate.R and renv/settings.dcf files.

Then run renv::restore() inside your application, instead of using the calls to remotes::install_version() as they are currently implemented when doing it with {golem}.

At the time of writing these lines, there is no native support of {renv} (with or without Docker) in {golem}, but that is something we can expect to happen in future versions of {golem}.

E. Develop inside a Docker container

Developers have their own R versions and operating systems, which generally differ from the one used on the production server, leading to issues when it comes to deploying the application.

If you plan on using Docker as a deployment mechanism, you can also use Docker as a local developer environment. Thanks to the containers maintained by the The Rocker Project[13], it's possible to have a local environment that comes close to what you will find on the production server. What's even more interesting is that this project offers images that can contain RStudio server: that means that the application that you will deploy in production can have the very same configuration as the one developers are using on their local machine: thanks to these containers, developers can work on a version of R that matches the one from the production server, using packages that will exactly match the one used in production.

Even more interesting is using RStudio inside Docker in combination with {renv}: the developers work on their machines, inside an IDE they know, and with system requirements (R versions, packages, etc.) that can be reproduced on the production server!

F. Read more about Docker

- An Introduction to Docker for R Users[14]

- An Introduction to Rocker: Docker Containers for R (Boettiger and Eddelbuettel, 2017)

- The Rockerverse: Packages and Applications for Containerization with R (Nüst et al., 2020)

---

[13]https://www.rocker-project.org/
[14]https://colinfay.me/docker-r-reproducibility/

12

12.1 Using version control with Git

"Friends do not let friends work on a coding project without version control." You might have heard this before, without really considering what this means. Or maybe you are convinced about this saying, but have not had the opportunity to use `Git`, GitHub or GitLab for versioning your applications. If so, now is the time to update your workflow!

12.1.1 Why version control?

Have you ever experienced a piece of code disappearing? Or the unsolvable problem of integrating changes when several people have been working on the same piece of code? Or the inability to find something you have written a while back?

If so, you might have been missing version control (also shortened as VC). In this chapter, we'll be focusing on `Git`, but you should be aware that other VC systems exist. As they are less popular than `Git`, we will not cover them here. `Git` was designed to handle collaboration on code projects [1] where potentially a lot of people have to interact and make changes to the codebase. `Git` might feel a little bit daunting at first, and even seasoned developers still misuse it, or do not understand it completely, but getting at ease with the basics will significantly improve the way you build software, so do not give up: the benefits from learning it really outweigh the (apparent) complexity.

There are many advantages to VC, including:

- **You can go back in time**. With a VC system like `Git`, every change is recorded (well, every **committed** change), meaning that you can potentially go back in time to a previous version of a project, and see the complete history of a file. This feature is very important: if you accidentally made changes that break your application, or if you deleted a feature you thought

---

[1] It was first developed by Linus Torvalds, the very same man behind Linux.

you would never need, you can go back to where you were a few hours, a few days, a few months back.

- **Several people can work on the same file**. Git relies on a system of branches. Within this branch pattern, there is one main branch, called "main", which contains the stable, main version of the code-base. By "forking" this branch (or any other branch), developers will have a copy of the base branch, where they can safely work on changing (and breaking) things, without impacting the origin branch. This allows you to try things in a safe environment, without touching what works. Note that simultaneously working on the same file at the same time might not be the perfect practice, it's better, if possible, to split the code into smaller files.

- **You can safely track changes**. Every time a developer records something to Git, changes are listed. In other words, you can see what changes are made to a specific file in your codebase.

- **It centralizes the codebase**. You can use Git locally, but its strength also relies on the ability to synchronize your local project with a distant server. This also means that *several* people can synchronize with this server and collaborate on a project. That way, changes on a branch on a server can be downloaded (it is called pull in Git terminology) by all the members of the team, and synchronized locally, i.e. if someone makes changes to a branch and sends them to the main server, all the other developers can retrieve these changes on their machine.

12.1.2   Git basics: add - commit - push - pull

These are the four main actions you will be performing in Git: if you just need to learn the minimum to get started, they are the four essential ones.

add

When using add, you are choosing which elements of your project you want to track, be it new files or modifications of an already versioned file. This action does not save the file in the Git repository, but flags the changes to be added to the next commit.

commit

A commit is a snapshot of a codebase at a given moment in time. Each commit is associated with two things: a sha1, which is a unique reference in the history of the project, allowing you to identify this precise state when you need to get back in time, and a message, which is a piece of text that describes the commit.[2] Note that messages are mandatory, you cannot commit without them, and that the `sha1` references are automatically generated by `Git`. Do not overlook these messages: they might seem like a constraint at first but they are a life saver when you need to understand the history of a project.

There is no strict rule about what and when to commit. Keep in mind that commits are what allow you to go back in time, so a commit is a complete state of your codebase to which it would make sense to return. A good practice is to state in the commit message which choices you made and why (but not how you implemented these changes), so that other developers (and you in the future) will be able to understand changes. Commit messages are also where you might specify the breaking changes, so that other developers can immediately see these when they are merging your code.

push

Once you have a set of commits ready, you are ready to **push** it to the server. In other words, you will permanently record these commits (hence the series of changes) to the server.

Making a push implies three things:

- Other people in the team will be able to retrieve the changes you have made.

- These changes will be recorded permanently in the project history.

- You cannot modify commits once they were sent to the server.[3]

pull

Once changes have been recorded in the main server, everybody synchronized with the project can **pull** the commits to their local project.

12.1.3 About branches

Branches are the `Git` way to organize work and ideas, notably when several people are collaborating on the same project (which might be the case when building large web applications with R).

How does it work? When you start a project, you are in the main branch, which is called the "main". In a perfect world, you never work directly on

---

[2]For example: "Added a graph in the analysis tab" or "Fixed the docx export bug".

[3]If you want to modify some code or have to go back in time, the best way to do it is to create a new commit with these changes or use adequate `Git` commands.

this branch: it should always contain a working, deployable version of the application.

Other branches are to be thought of as work areas, where developers fix bugs or add features. The modifications made in these development branches will then be transferred (directly or indirectly) to the main branch. This principle is shown in Figure 12.1.

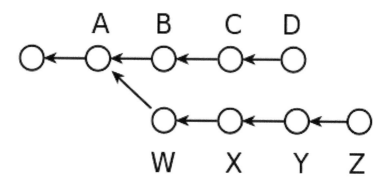

FIGURE 12.1: Branches in `Git`.

In practice, you might want to use a workflow where each branch is designed to fix a small issue or implement a feature, so that it is easier to separate each small part of the work. Even when working alone.

12.1.4   Issues

If you are working with a remote tool with a graphical interface like GitLab, GitHub or Bitbucket, there is a good chance you will be using issues. Issues are "notes" or "tickets" that can be used to track a bug or to suggest a feature. This tool is crucial when it comes to project management: issues are the perfect spot for organizing and discussing ideas, but also to have an overview of what has been done, what is currently being done, and what is left to be done. Issue may also be used as a discussion medium with beta testers, clients or sponsors.

One other valuable feature of issues is that they can be referenced inside commits using a hashtag and its number: #123. In other words, when you send code to the centralized server, you can link this code to one or more issues and corresponding commits appear in the issue discussions.

12.2 Git integration

12.2.1 With RStudio

`Git` is very well integrated in the RStudio IDE, and using `Git` can be as simple as clicking on a button from time to time. If you are using RStudio, you will find a pull/push button, a stage and commit interface, and a tool for visualizing differences in files. Everything you need to get started is there.

Note that of course, it will be better in the long run to get a more complete understanding of how `Git` works, so that when things get more complexe, you will be able to handle them.

12.2.2 As part of a larger world

`Git` is not reserved for team work: even if you are working alone on a project, using `Git` is definitely worth the effort. Using `Git`, and particularly issues, helps you organize your train of thought, especially upfront when you need to plan what you will be doing.

And of course, remember that `Git` is not limited to `{shiny}` applications: it can be used for any other R-related projects, and at the end of the day for any code related projects, making it a valuable skill to have in your toolbox, whatever language you will be working with in 10 years!

12.2.3 About git-flow

There are a lot of different ways and methodologies to organize your `Git` workflow. One of the most popular ones is called `git flow`, and we will give you here a quick introduction on how you can manage your work using this approach. Please note that this is a quick introduction, not a complete guide: we will link to some further reading just at the end of this section.

So, here are the key concepts of `git flow`:

- The `main` branch only contains stable code: most of the time it matches a tagged, fixed version (v0.0.1, 0.1.0, v1.0.0, etc.). A very small subset of developers involved in the project have writing access to the `main` branch, and no developer should ever push code straight to this branch: new code to `main` only comes either from the `dev` branch, or from a `hotfix` branch. For an app in production, the last commit of this branch should be the version that is currently in production.

- The `dev` branch, on the other hand, is the "Work in progress" branch: the one that contains the latest changes before they are merged into main. This is the common working branch for every developer. Most of the time, developers do not push code into these branches either: they make merge/pull requests (MR/PR) to `dev` from a `feature branch`.

- A `feature branch` is one branch, forked from `dev`, that implements one of the features of the application. To keep a clean track of what each branch is doing, a good practice is to use `issue-XXX`, where `XXX` is the corresponding issue you plan to solve in this branch.

- A `hot fix` branch is a branch to correct a critical issue in `main`. It is forked from `main`, and is merged straight into `main` using an MR.

A summary of this process is available in Figure 12.2.

From a software engineer point of view, here is how daily work goes:

- Identify an issue to work on.

- Fork dev into `issue-XXX`.

- Develop a feature inside the branch.

- Regularly run `git stash`, `git rebase dev`, and `git stash apply` to include the latest changes from `dev` to stay synchronized with `dev`.[4]

- Make a pull request to `dev` so that the feature is included.

- Once the PR is accepted by the project manager, notify the rest of the team that there have been changes to `dev`, so they can rebase it to the branch they are working on.

- Start working on a new feature.

Of course, there are way more subtleties to this flow of work, but this gives you a good starting point. Generally speaking, good communication between developers is essential for a successful collaborative development project.

---

[4]There are two strategies for merging `dev`: either a "merge strategy" or a "rebase strategy". Both strategies have pros and cons. We work with the "rebase strategy" to force ourselves to stay updated. We also notice that this strategy lowers the risk of bad merging, that can cause code loss. However, this requires a lot of communication between developers and a good knowledge of `Git`.

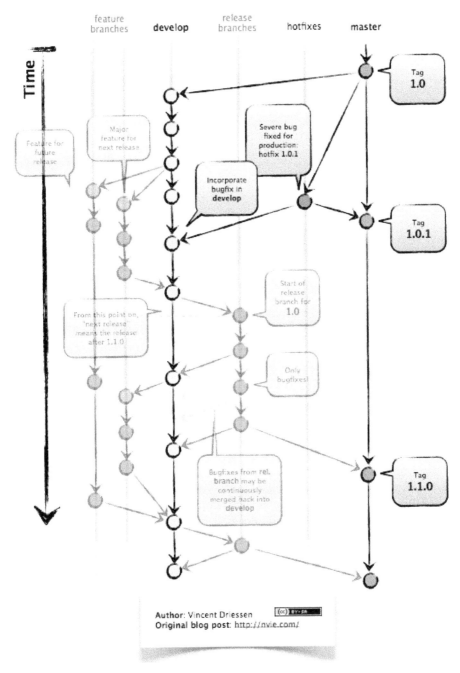

FIGURE 12.2: Presentation of a `git flow` (Vincent Driessen, http://nvie.com).

12.2.4   Further readings on `Git`

If you want to learn more about `Git`, here are some resources that have helped us in the past:

- `https://happygitwithr.com/`
- `https://git-scm.com/book`
- `https://www.git-tower.com/blog/git-cheat-sheet/`

12.3   Automated testing

We have seen in Chapter 11 how to build a testing infrastructure for your app, notably using the `{testthat}` (Wickham, 2021) package. What we have described is a way to build it locally, before running your test on your own machine. But there is a big flaw to this approach: you have to remember to run the tests, be it regularly or before making a pull request/pushing to the server. To do this kind of job, you will be looking for **a tool to do automated testing at the repository level: in other words, a software that can test your application whenever a piece of code is pushed/moved on the repository**.

To do this, various tools are available, each with their own features. Here is a non-exhaustive list of the ones you can choose:

Travis CI[5] is a software that can be synced with your `Git` repositories (GitHub or Bitbucket), and whenever something happens on the repo, the events described in the travis configuration file (`.travis.yml`) are executed. If they exit with a code 0, the test passes. If they do not, the integrated tests have failed. Travis CI integration may be used internally and externally: internally, in the sense that before merging any pull request, the project manager has access to a series of tests that are automatically launched. Externally, as a "health check" before installing software: if you visit a GitHub repository that has Travis badges included, you can check if the current state of the package/software is stable, *i.e.* if it passes the automated tests.

Travis CI can do a lot more than just testing your app: it can be used to build documentation, deploy to production, or to run any other scripts you want to be run before/after the tests have passed. And the nice thing is that you can test for various versions of R, so that you are sure that you are supporting current, future and previous versions of R.

---
[5]`https://travis-ci.org/`

All of this is defined in the `.travis.yml` file, which is to be put at the root of your source directory, a file that is automatically generated when calling `usethis::use_travis()`.

Note that Travis CI can run tests on GNU/Linux or MacOS operating systems.

Appveyor[6] has the same functionalities as Travis CI. This service can integrate with GitHub, GitHub Enterprise, Bitbucket, GitLab, Azure Repos, Kiln, Gitea. It supports Windows, Linux and macOS.

GitHub actions[7] serve a related purpose: defining actions to be performed as responses to events on the GitHub repository. Testing, building documentation, push to another repository, deploy on the server—all these actions can be automatically performed. As with Travis CI, these actions are defined in a `yaml` file. Examples of these configurations can be found at r-lib/actions[8], and some can be automatically linked to your project using functions from {usethis}: `use_github_action_check_release()`, `use_github_action_check_standard()`, `use_github_action_check_full()` and `use_github_action_pr_commands()`. The first three perform a standard `R CMD check`, under various conditions:

- The `release` tests on MacOS, with the latest version of R, and runs the check via the {rcmdcheck} (Csárdi, 2019b) package.
- `standard` does the check for 3 operating systems (Windows, Mac and Linux), and for R and R-devel.
- `full` does `standard` but for the last 5 minor versions of R.

Finally, `use_github_action_pr_commands()` sets checks to be performed when a pull request is made to the repository.

If you are working with GitLab, you can use the integrated `GitLab CI` service: it serves the same purpose, with the little difference that it is completely Docker-based: you define a `yaml` with a series of stages that are performed (concurrently or sequentially), and they are all launched inside a Docker container. To help you with this, the `colinfay/r-ci-tidyverse`[9] Docker image comes with pre-installed packages for testing: {remotes} (Hester et al., 2021), {testthat} (Wickham, 2021), {config} (Allaire, 2020) and is available for several R versions. This Docker image can be used as the source image for your `GitLab CI` yaml file.

Here is an example of one of these files:

---

[6]https://www.appveyor.com/
[7]https://github.com/features/actions
[8]https://github.com/r-lib/actions
[9]https://hub.docker.com/r/colinfay/r-ci-tidyverse

```yaml
image: colinfay/r-ci-tidyverse:3.6.0

cache:
  paths:
    - ci/

stages:
  - test
  - document

building:
  stage: test
  script:
    - R -e "remotes::install_deps(dependencies = TRUE)"
    - R -e 'devtools::check()'

documenting:
    stage: document
    allow_failure: true
    when: on_success
    only:
        - main
    script:
        - Rscript -e 'install.packages("DT")'
        - Rscript -e 'covr::gitlab(quiet = FALSE)'
    artifacts:
        paths:
            - public
```

Automated testing, continuous integration, and continuous deployment are vast topics that cannot be covered in a few pages inside this book, but spending some time learning about these methodologies is definitely worth the time spent: the more you can automate these processes, and the more you test, the more your application will be resilient, easy to maintain, and easy to enhance: the more you check, the quicker you will discover bugs.

And the quicker you detect bugs, the easier it is to correct them!

Part VI

Step 5: Deploy

13

Deploy Your Application

Your deploys should be as boring, straightforward, and stress-free as possible.

How to Deploy Software - Zach Holman (`https://zachholman.com/posts/deploying-software`)

Once your app is built, you are ready to deploy it! In other words, your software is now ready to be used by other users. There are two main ways to share your application and make it available to others: by creating a package and making it installable, or by sending it to a remote server. We will see in this part how you can do that using {golem} (Fay et al., 2021).

13.1 Before deployment checklist

Here is a quick checklist of things to think about once your application is ready, and before sending it to production:

- ☐ `devtools::check()`, run from the command line, returns 0 errors, 0 warnings, and 0 notes.

- ☐ The current version number is valid, i.e. if the current app is an update, the version number has been bumped.

- ☐ Everything is fully documented.

- ☐ Test coverage is good, i.e. you cover a sufficient amount of the codebase, and these tests cover the core/strategic algorithms

☐ Everyone in the project knows the person to call if something goes wrong.

☐ The following things are clear to everyone involved in the project: the debugging process, how to communicate bugs to the developer team, and how long it will take to get changes implemented.

☐ (If relevant) The server it is deployed on has all the necessary software installed (Docker, Connect, `Shiny Server`, etc.) to make the application run.

☐ The server has all the system requirements needed (i.e. the system libraries), and if not, they are installed with the application (if it's dockerized).

☐ The application, if deployed on a server, will be deployed on a port which will be accessible by the users.

☐ (If relevant) The environment variables from the production server are managed inside the application.

☐ (If relevant) The app is launched on the correct port, or at least this port can be configured via an environment variable.

☐ (If relevant) The server where the app is deployed has access to the data sources (database, API, etc.).

☐ If the app records data, there are backups for these data.

13.2 Sharing your app as a package

13.2.1 Install on your machine

A {shiny} application built with {golem} (Fay et al., 2021) is **by definition** an R package. This {shiny} app as a package is also helpful when it comes to deploying your application: packages are designed to be shareable pieces of R code.

Before sending it to a remote server or sharing it with the world, **the first step is testing if the package can be installed on your own computer.** To do that, when you are in the project corresponding to the golem you built, you can call `remotes::install_local()` to install the application on your computer. Of course, if you are somewhere else on your machine, you can call `remotes::install_local("path/to/app")`. If you are using the RStudio IDE, you can also click on the `Build` tab, then click on the `Install and Restart` button.

This should restart your R session, and call `library(yourpackagename)`. Then, try the `run_app()` function to check that the app can be launched.

13.2.2 Share as a built package

A. Local build

Building an app as a package also means that this app can be bundled into an archive, and then shared, either as is or using a package repository like the CRAN.

To do that, you first need an bundled version of your app, which can be created using the `build()` function from {pkgbuild} (Wickham and Hester, 2020) in the same working directory as your application. Calling this function will create a .tar.gz file that is called `mygolem_0.0.1.tar.gz` (of course with the name of your package). Once you have this `tar.gz`, you can send it to your favorite package repository.

You can also share the file as is with others. If you do so, they will have to install the app with `remotes::install_local("path/to/tar.gz")`, that will take care of doing a full installation of the app, including installing the required dependencies. Then, they can do `library(yourpackagename)` and `run_app()` on their machine.

B. Send to a package repository

The upside of building the application {golem}, i.e. as a package, is that you can share your application on a remote package manager, the more widely used, for example, on the CRAN like {dccvalidator} (Woo et al., 2020), or on BioConductor like {spatialLIBD} (Collado-Torres et al., 2020). But any other package manager will work: for example, if the company uses RStudio Package Manager, your application can be installed here in the same way as any other package. If your application is open source, the package structure also allows you to install from GitHub, by using the `remotes::install_github()` function.[1] For example, this is what you can do with {hexmake} or {tidytuesday}: as they are open-source packages, they can be installed from GitHub. Then, once your application is installed as a package on the users' machines, they can do `library(yourpackagename)` and `run_app()`.

The advantage of this solution is that R users are familiar with package installation, so it makes using your application easier for them. Also, and we will see it in the next section, making your application available as a standard

---

[1]This is also true for other version control systems.

R package makes it easier to deploy it: for example, if your RStudio Connect is coupled with your RStudio Package Manager, the deployment file just has to contain one line launching the application.

Note that releasing to CRAN or BioConductor requires extra effort: you have to comply with a series of rules. But good news: as you have been following the best practices from this book, you should not have to put in that much extra effort!

Know more about releasing on CRAN:

- Checklist for CRAN submissions[2]
- CRAN Repository Policy[3]
- R packages - Chapter 18, Releasing a package[4]
- Getting your R package on CRAN[5]
- prepare-for-cran - A collaborative list of things to know before submitting to CRAN[6]

13.3   Deploying apps with {golem}

The other way to make your application available to others is by sending it to a remote server that can serve {shiny} applications. In other words, instead of having to install the application on their machines, **they can crack open a web browser and navigate to the URL where the application is deployed**. Deploying to a server is the solution of choice when you want to make your application available to a wide public: on a server, visitors do not have to have R installed on their computer, they do not have to install a package or launch it; they can just browse the application like any other web application. This solution is also a common choice in companies that have strict security requirements: the IT team might not be willing to let everyone install software on their machine, and sharing an application on a server allows them more control over who can access the application. For example, deploying on a server allows you to use a proxy, and to filter by IP: then, only a subset of people can have access to the application.

When using {golem}, you can open the dev/03_deploy.R and find the functions for server deployment. At the time of writing this book, there are two main ways to deploy a shiny app on a server:

---

[2]https://cran.r-project.org/web/packages/submission_checklist.html
[3]https://cran.r-project.org/web/packages/policies.html
[4]https://r-pkgs.org/release.html
[5]https://kbroman.org/pkg_primer/pages/cran.html
[6]https://github.com/ThinkR-open/prepare-for-cran

- RStudio's solutions
- A Docker-based solution

13.3.1 RStudio environments

RStudio proposes three services to deploy {shiny} application:

- shinyapps.io, an on-premises solution, can serve {shiny} application (freemium).
- Shiny Server is a software you have to install on your own server, and can be used to deploy multiple applications (you can find either an open source or a professional edition).
- RStudio Connect is a server-based solution that can deploy {shiny} applications and Markdown documents (and other kinds of content), and serves them as ordinary websites.

Each of these platforms has its own function to create an app.R file that is to be used as a launch script of each platform.

- golem::add_rstudioconnect_file()

- golem::add_shinyappsio_file()

- golem::add_shinyserver_file()

These app.R files call a pkgload::load_all() function, that will mimic the launch of your package, and then call the run_app() function from your packaged app. Note that if you need to configure the way your app is launched on these platforms (for example, if you need to pass arguments to the run_app() function), you will have to edit this file.

Note that when using these functions, you will be able to use the "One click deploy" for these platforms: on the top right of these app.R, use the Blue Button to deploy to a server.

Another way to deploy your {golem}-based app to {shiny} server and to Connect is to link these two software to a local repository (for example, an RStudio Package Manager), and then to only use mypackage::run_app() to the app.R.

13.3.2 Docker

Docker is an open source software used to build and deploy applications in containers. Docker has become a core solution in the DevOps world and a lot of server solutions are based on it. See Part 5, "Strengthen", for a more complete introduction to Docker.

You will find the function for creating a `Dockerfile` for your {golem} app inside the `03_deploy.R` file, which contains a series of 3 functions:

- `golem::add_dockerfile()`
- `golem::add_dockerfile_shinyproxy()`
- `golem::add_dockerfile_heroku()`

The first function creates a "generic" `Dockerfile`, in the sense that it is not specific to any platform, and would work out of the box for your local machine. The second one is meant for {shiny}Proxy[7], an open source solution for deploying containerized {shiny} applications, and the third is for Heroku[8], an online service that can serve containerized applications (not specific to {shiny}).

Other platforms can run Docker containers, notably AWS and Google Cloud Engine. At the time of writing these lines, {golem} does not provide support for these environments, but that is on the to-do list!

Note that the `Dockerfile` creation in {golem} tries to replicate your local environment as precisely as possible, notably by matching your R version, and the version of the packages you have installed on your machine. System requirements are also added when they are found on the sysreqs service from r-hub[9]. Otherwise you might have to add them manually.

---

[7]https://www.shinyproxy.io/
[8]https://www.heroku.com/
[9]https://sysreqs.r-hub.io/

Part VII

Optimizing

14

The Need for Optimization

Only once we have a solid characterization of the surface area we want to improve can we begin to identify the best way to improve it.

Refactoring at Scale (Lemaire, 2020)

14.1 Build first, then optimize

14.1.1 Identifying bottlenecks

Refactoring existing code for speed sounds like an appealing activity for a lot of us: it is always satisfying to watch our functions get faster, or finding a more elegant way to solve a problem that also results in making your code a little bit faster. Or as Maude Lemaire writes in *Refactoring at Scale* (Lemaire, 2020), "Refactoring can be a little bit like eating brownies: the first few bites are delicious, making it easy to get carried away and accidentally eat an entire dozen. When you've taken your last bite, a bit of regret and perhaps a twinge of nausea kick in."

But beware! As Donald Knuth puts it "Premature optimization is the root of all evil". What does that mean? That **focusing on optimizing small portions of your app before making it work fully is the best way to lose time along the way, even more in the context of a production application, where there are deadlines and a limited amount of time to build the application**. Why? Here is the general idea: let's say the schema in Figure 14.1 represents your software, and its goal is to make things travel from $X1$ to $X2$, but you have a bottleneck at U. You are building elements piece by piece: first, the portion X1.1 of the "road", then X1.2, etc. Only

215

when you have your application ready can you really appreciate where your bottleneck is, and you can focus on making things go fast from X1.1 to X.1.2, these performance gains won't make your application go faster: you will only make the elements move faster to the bottleneck.

When? Once the application is ready: here in our example, we can only detect the bottleneck once the full road is actually built, not while we are building the circle.

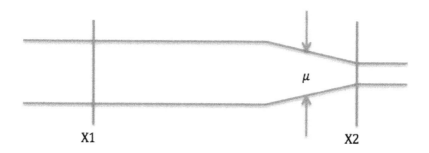

FIGURE 14.1: Road bottleneck, from WikiMedia `https://commons.wikimedia.org/wiki/File:Roadway_section_with_bottleneck.png`.

This bottleneck is the very thing you should be optimizing: **having faster code anywhere else except this bottleneck will not make your app faster**: you will just make your app reach the bottleneck faster, but there will still be this part of your app that slows everything down. But this is something you might only realize when the app is fully built: pieces might be fast individually, but slow when put together. It is also possible that the test dataset you have been using from the start works just fine, but when you try your app with a bigger, more realistic dataset, the application is actually way slower than it should be. And, maybe you have been using an example dataset so that you do not have to query the database every time you implement a new feature, but the SQL query to the database is actually very slow. This is something you will discover only when the application is fully functional, not when building the parts, and realizing that when you only have 5% of the allocated time for this project left on your calendar is not a good surprise.

Or to sum up:

Get your design right with an un-optimized, slow, memory-intensive implementation before you try to tune. Then, tune systematically, looking for the places where you can buy big performance wins with the smallest possible increases in local complexity.

The Art of UNIX Programming (Raymond, 2003)

14.1.2 Do you need faster functions?

Optimizing an app is a matter of trade-offs: of course, in a perfect world, every piece of the app would be tailored to be fast, easy to maintain, and elegant. But in the real world, you have deadlines, limited time and resources, and we are all but humans. That means that at the end of the day, your app will not be completely perfect: software can **always** be made better. No piece of code has ever reached complete perfection.

Given that, **do you want to spend 5 days out of the 30 you have planned optimizing a function so that it runs in a quarter of a second instead of half a second**, then realize the critical bottleneck of your app is actually the SQL query and not the data manipulation? Of course a function running twice as fast is a good thing, but think about it in context: for example, how many times is this function called? We can safely bet that if your function is only called once, working on making it twice as fast might not be the one function you would want to focus on (well, unless you have unlimited time to work on your project, and in that case lucky you; you can spend a massive amount of time building the perfect software). On the other hand, the function which is called thousands of times in your application might benefit from being optimized.

And all of this is basic maths. Let's assume the following:

- A current scenario takes 300 seconds to be accomplished on your application.
- One function A() takes 30 seconds, and it's called once.
- One function B() takes 1 second, and it's called 50 times.

If you divide the execution time of A() by two, you would be performing a local optimization of 15 seconds, and a global optimization of 15 seconds. On the other hand, if you divide the execution time of B() by two, you would be performing a local optimization of 0.5 seconds, but a global optimization of 25 seconds.

Again, this kind of optimization is hard to detect until the app is functional. An optimization of 15 seconds is way greater than an optimization of 0.5 seconds. Yet you will only realize that once the application is up and running!

14.1.3 Don't sacrifice readability

As said in the last section, every piece of code can be rewritten to be faster, either from R to R or using a lower-level language: for example C or C++. You can also rebuild data manipulation code switching from one package to another, or use a complex data structures to optimize memory usage, etc.

But that comes with a price: **not keeping things simple for the sake of local optimization makes maintenance harder, even more if you are using a lesser-known language/package**. Refactoring a piece of code is better done when you keep in mind that "the primary goal should be to produce human-friendly code, even at the cost of your original design. If the laser focus is on the solution rather than the process, there's a greater chance your application will end up more contrived and complicated than it was in the first place" (Lemaire, 2020).

For example, switching some portions of your code to C++ implies that you might be the only person who can maintain that specific portion of code, or that your colleague taking over the project will have to spend hours learning the tools you have been building, or the language you have chosen to write your functions with.

Again, **optimization is always a matter of trade-off**: is the half-second local optimization worth the extra hours you will have to spend correcting bugs when the app will crash and when you will be the only one able to correct it? Also, are the extra hours/days spent rewriting a working code-base worth the speed gain of 0.5 seconds on one function?

For example, let's compare both these implementations of the same function, one in R, and one in C++ via {Rcpp} (Eddelbuettel et al., 2021). Of course, the C++ function is faster than the R one—this is the very reason for using C++ with R.

```
library("Rcpp")
# A C++ function to compute the mean
cppFunction("
double mean_cpp(NumericVector x) {
  int j;
  int size = x.size();
  double res = 0;
  for (j = 0; j < size; j++){
```

```
    res = res + x[j];
  }
  return res / size;
}")

# Computing the mean using base R and C++,
# and comparing the time spent on each
benched <- bench::mark(
  cpp = mean_cpp(1:100000),
  native = mean(1:100000),
  iterations = 1000
)
benched
```

```
# A tibble: 2 x 6
  expression     min median `itr/sec` mem_alloc `gc/sec`
  <bch:expr> <bch:> <bch:>     <dbl> <bch:byt>    <dbl>
1 cpp          115us  496us     1627.     784KB     6.53
2 native       520us  652us     1211.       0B        0
```

(Note: we will come back to `bench::mark()` later.)

However, how much is a time gain worth if you are not sure you can get someone on your team to take over the maintenance if needed? In other words, given that (in our example) we are gaining around -1.5631×10^{-4} on the execution time of our function, is it worth switching to C++? Using external languages or complex data structures implies that from the start, you will need to think about who and how your codebase will be maintained over the years.

Chances are that if you plan on using a {shiny} application during a span of several years, various R developers will be working on the project, and including C++ code inside your application means that these future developers will either be required to know C++, or they will not be able to maintain this piece of code.

So, to sum up, there are three ways to optimize your application and R code, and the bad news is that you cannot optimize for all of them:

- Optimizing for speed
- Optimizing for memory
- Optimizing for readability/maintainability

Leading a successful project means that you should, as much as possible, find the perfect balance between these three.

14.2   Tools for profiling

14.2.1   Profiling R code

A. Identifying bottlenecks

The best way to profile R code is by using the {profvis} (Chang et al., 2020) package,[1] a package designed to evaluate how much time each part of a function call takes. With {profvis}, you can spot the bottleneck in your function. Without an automated tool to do the profiling, the developers would have to profile by guessing, which will, most of the time, come with bad results:

One of the lessons that the original Unix programmers learned early is that intuition is a poor guide to where the bottlenecks are, even for one who knows the code in question intimately.

The Art of UNIX Programming (Raymond, 2003)

Instead of guessing, it is a safe bet to go for a tool like {profvis}, which allows you to have a detailed view of what takes a long time to run in your R code.

Using this package is quite straightforward: put the code you want to benchmark inside the profvis() function,[2] wait for the code to run, and that is it; you now have an analysis of your code running time.

Here is an example with 3 nested functions, top(), middle() and bottom(), where top() calls middle() which calls bottom():

```r
library(profvis)
top <- function(){
  # We use profvis::pause() because Sys.sleep() doesn't
  # show in the flame graph
```

---

[1]{utils} also comes with a function call Rprof(), but we will not be examining this one here, as {profvis} provides a more user-friendly and enhanced interface to this profiling function.

[2]Do not forget to add {} inside profvis({}) if you want to write several lines of code.

```
  pause(0.1)
  # Running a series of function with lapply()
  lapply(1:10, function(x){
    x * 10
  })
  # Calling a lower level function
  middle()
}

middle <- function(){
  # Pausing before computing, and calling other functions
  pause(0.2)
  1e4 * 9
  bottom_a()
  bottom_b()
}

# Both will pause and print, _a for 0.5 seconds,
# _b for 2 seconds
bottom_a <- function(){
  pause(0.5)
  print("hey")
}
bottom_b <- function(){
  pause(2)
  print("hey")
}
profvis({
  top()
})
```

What you see now is called a **flame graph**: it is a detailed timing of how your function has run, with a clear decomposition of the call stack. What you see in the top window is the expression evaluated, and on the bottom the details of the call stack, with what looks a little bit like a Gantt diagram. This result reads as follow: the wider the function call, the more time it has taken R to compute this piece of code. On the very bottom, the "top" function (i.e. the function which is directly called in the console), and the higher you go, the more you enter the nested function calls.

Here is how to read the graph in 14.2:

- On the x axis is the time spent computing the whole function. Our top() function being the only one executed, it takes the whole record time.

- Then, the second line shows the functions which are called inside `top()`. First, R pauses, then does a series of calls to `FUN` (which is the internal anonymous function from `lapply()`), and then calls the `middle()` function.

- Then, the third line details the calls made by `middle()`, which pauses, then calls `bottom_a()` and `bottom_b()`, which each `pause()` for a given amount of time.

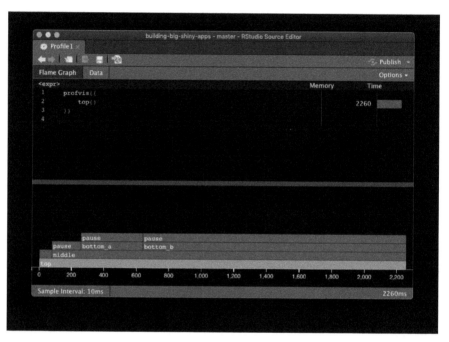

FIGURE 14.2: {profvis} flame graph.

If you click on the "Data" tab, you will also find another view of the **flame graph**, shown in 14.3, where you can read the hierarchy of calls and the time and memory spent on each function call:

If you are working on profiling the memory usage, you can also use the {profmem} (Bengtsson, 2020b) package which, instead of focusing on execution time, will record the memory usage of calls.

```
library(profmem)
# Computing the memory used by each c
p <- profmem({
  x <- raw(1000)
  A <- matrix(rnorm(100), ncol = 10)
```

FIGURE 14.3: {profvis} data tab.

```
})
p
```

```
Rprofmem memory profiling of:
{
    x <- raw(1000)
    A <- matrix(rnorm(100), ncol = 10)
}

Memory allocations:
      what bytes                  calls
1    alloc  1048                  raw()
2    alloc   272               matrix()
3    alloc   560               matrix()
4    alloc   552               matrix()
5    alloc  1072               matrix()
6    alloc   848 matrix() -> rnorm()
7    alloc  2552 matrix() -> rnorm()
8    alloc   848               matrix()
9    alloc   528            <internal>
```

```
10      alloc   1648            <internal>
11      alloc   1648            <internal>
12      alloc   1072            <internal>
13      alloc   256             <internal>
14      alloc   456             <internal>
15      alloc   216             <internal>
16      alloc   256             <internal>
total           13832
```

You can also get the total allocated memory with:

```
total(p)
```

```
[1] 13832
```

And extract specific values based on the memory allocation:

```
p2 <- subset(p, bytes > 1000)
print(p2)
```

```
Rprofmem memory profiling of:
{
    x <- raw(1000)
    A <- matrix(rnorm(100), ncol = 10)
}

Memory allocations:
        what bytes                  calls
1       alloc   1048                 raw()
5       alloc   1072               matrix()
7       alloc   2552 matrix() -> rnorm()
10      alloc   1648            <internal>
11      alloc   1648            <internal>
12      alloc   1072            <internal>
total           9040
```

(Example extracted from {profmem} help page).

Here it is; now you have a tool to identify bottlenecks!

B. Benchmarking R code

Identifying bottlenecks is a start, but what to do now? In the next chapter about optimization, we will dive deeper into common strategies for optimizing R and {shiny} code. But before that, remember this rule: **never start optimizing if you cannot benchmark this optimization**. Why? Because developers are not perfect at identifying bottlenecks and estimating if something is faster or not, and some optimization methods might lead to slower code. Of course, most of the time they will not, but in some cases adopting optimization methods leads to writing slower code, because we have missed a bottleneck in our new code. And of course, without a clear documentation of what we are doing, we will be missing it, relying only on our intuition as a rough guess of speed gain.

In other words, if you want to be sure that you are actually optimizing, be sure that you have a basis for comparison.

How to do that? One thing that can be done is to keep an RMarkdown file with your starting point: use this notebook to keep track of what you are doing, by noting where you are starting from (i.e, what's the original function you want to optimize), and compare it with the new one. By using an Rmd, you can document the strategies you have been using to optimize the code, e.ga: "switched from for loop to vectorize function", "changed from x to y", etc. This will also be helpful for the future: either for you in other projects (you can get back to this document), or for other developers, as it will explain why specific decisions have been made.

To do the timing computation, you can use the {bench} (Hester, 2020a) package, which compares the execution time (and other metrics) of two functions. This function takes a series of named elements, each containing an R expression that will be timed. Note that by default, the mark() function compares the output of each function,

Once the timing is done, you will get a data.frame with various metrics about the benchmark.

```r
# Multiplying each element of a vector going from 1 to size
# with a for loop
for_loop <- function(size){
  res <- numeric(size)
  for (i in 1:size){
    res[i] <- i * 10
  }
  return(res)
}
# Doing the same thing using a vectorized function
```

```
vectorized <- function(size){
  (1:size) * 10
}
res <- bench::mark(
  for_loop = for_loop(1000),
  vectorized = vectorized(1000),
  iterations = 1000
)
res
```

```
# A tibble: 2 x 6
  expression       min    median `itr/sec` mem_alloc
  <bch:expr> <bch:tm> <bch:tm>      <dbl> <bch:byt>
1 for_loop    46.67us   81.2us      9540.    30.5KB
2 vectorized   5.85us   10.9us      4356.    11.8KB
# ... with 1 more variable: gc/sec <dbl>
```

Here, we have an empirical evidence that one code is faster than the other: by benchmarking the speed of our code, we are able to determine which function is the fastest.

If you want a graphical analysis, {bench} comes with an `autoplot` method for {ggplot2} (Wickham et al., 2021a), as shown in Figure 14.4:

```
ggplot2::autoplot(res)
```

And, bonus point, {bench} takes time to check that the two outputs are the same, so that you are sure you are comparing the very same thing, which is another crucial aspect of benchmarking: be sure you are not comparing apples with oranges!

14.2.2 Profiling {shiny}

A. {shiny} back-end

You can profile {shiny} applications using the {profvis} package, just as any other piece of R code. The only thing to note is that if you want to use this function with an app built with {golem} (Fay et al., 2021), you will have to wrap the `run_app()` function in a `print()` function. Long story short, what makes the app run is not the function itself, but the printing of the function, so

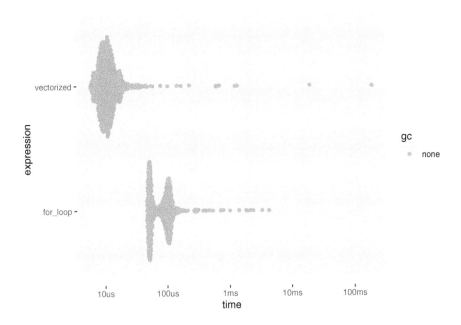

FIGURE 14.4: {bench} autoplot.

the object returned by `run_app()` itself cannot be profiled. See the discussion of this issue on the {golem} repository[3] to learn more about this.

B. {shiny} front-end

Google Lighthouse

One other thing that can be optimized when it comes to the user interface is the web page rendering performance. To do that, we can use standard web development tools: as said several times, a {shiny} application IS a web application, so tools that are language agnostic will work with {shiny}. There are thousands of tools available to do exactly that, and going through all of them would probably not make a lot of sense.

Let's focus on getting started with a basic but powerful tool, that comes for free inside your browser: Google Lighthouse[4], one of the famous tools for profiling web pages, is bundled into recent versions of Google Chrome. The nice thing is that this tool not only covers what you see (i.e. not only what you are actually rendering on your personal computer), but can also audit your app with various configurations, notably on mobile, with low bandwidth

---

[3]https://github.com/ThinkR-open/golem/issues/146
[4]https://developers.google.com/web/tools/lighthouse

and/or mimicking a 3G connection. **Being able to perform an audit of our application as seen on a mobile device is a real strength: we are developing an application on our computer, and might not be regularly checking how our application is performing on a mobile. Yet a large portion of web navigation is performed on a mobile or tablet.**

Already in 2016, Google wrote[5] that *"More than half of all web traffic now comes from smartphones and tablets"*. Knowing the exact number of visitors that browse through mobile is hard: the web is vast, and not all websites record the traffic they receive. Yet many, if not all, studies of how the web is browsed report the same results: more traffic is created via mobile than via computer.[6]

And, the advantages of running it in your browser is that it can perform the analysis on locally deployed applications: in other words, you can launch your {shiny} application in your R console, open the app in Google Chrome, and run the audit. A lot of online services need a URL to do the audit!

Each result from the audit comes with advice and changes you can make to your application to make it better, with links to know more about the specific issue.

And of course, last but not least, you also get the results of the metrics you have "passed". It is always a good mood booster to see our app passing some audited points!

Here is a quick introduction to this tool:

- Open Chrome in incognito mode (File > New Icognito Window),[7] so that the page performance is not influenced by any of the installed extensions in your Google Chrome.
- Open your developer console, either by going to View > Developer > Developer tools, by right-clicking > Inspect, or with the keyboard shortcut ctrl/cmd + alt + I, as shown in Figure 14.5.
- Go to the "Audit" tab.
- Configure your report (or leave the default).
- Click on "Generate Report".

Note that you can also install a command-line tool with `npm install -g lighthouse`,[8] then run `lighthouse http://urlto.audit`: it will produce either a JSON (if asked) or an HTML report (the default).

---

[5]https://www.thinkwithgoogle.com/data/web-traffic-from-smartphones-and-tablets/

[6]broadbandsearch https://www.broadbandsearch.net/blog/mobile-desktop-internet-usage-statistics for example, reports a 53.3% share for mobile browsing.

[7]This mode opens an "anonymous" session, in the sense that you don't have access to your account and extensions, and that the visits will not be recorded in your history.

[8]Being a NodeJS application, you will need to have NodeJS installed on your machine.

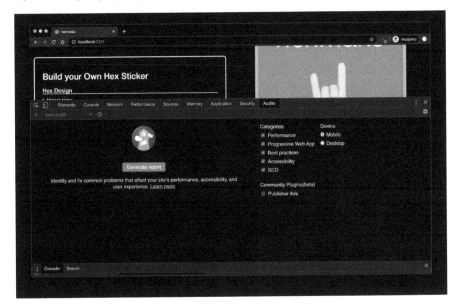

FIGURE 14.5: Launching Lighthouse audit from Google Chrome.

See Figure 14.6 for a screenshot of the results computed by Google Lighthouse.

FIGURE 14.6: Lighthouse audit results.

Once the audit is finished, you have some basic but useful indications about your application:

- Performance. This metric mostly analyzes the rendering time of the page: for example, how much time does it take to load the app in full, that is to say how much time it takes from the first byte received to the app being fully ready to be used, the time between the very first call to the server and the very first response, etc. With `{shiny}` (Chang et al., 2021a), you should get low performance here, notably due to the fact that it is serving external dependencies that you might not be able to control. For example, the report from `{hexmake}` (Fay, 2021f) suggests to "Eliminate render-blocking resources", and most of them are not controlled by the shiny developer: they come bundled with `shiny::fluidPage()` itself.

- Accessibility. Google Lighthouse performs a series of accessibility tests (see our chapter about accessibility for more information).

- Best practices bundles a list of "misc" best practices around web applications.

- SEO, search engine optimization, or how your app will perform when it comes to search engine indexation.[9]

- Progressive Web App (PWA): A PWA is an app that can run on any device, *"reaching anyone, anywhere, on any device with a single codebase"*. Google audit your application to see if your application fits with this idea.

Profiling web page is a wide topic and a lot of things can be done to enhance the global page performance. That being said, if you have a limited time to invest in optimizing the front-end performance of the application, Google Lighthouse is a perfect tool, and can be your go-to audit tool for your application.

And if you want to do it from R, the npm lighthouse module allows you to output the audit in JSON, which can then be brought back to R!

```
lighthouse --output json \
  --output-path data-raw/output.json \
  http://localhost:2811
```

Then, being a JSON file, you can call if from R:

---

[9]Search engine indexation refers to how Google ranks your website in the search results for a given query.

```
# Reading the JSON output of your lighthouse audit,
# and displaying the Speed Index value
lighthouse_report <- jsonlite::read_json("data-raw/output.json")
lighthouse_report$audits$`speed-index`$displayValue
```

```
[1] "5.4 s"
```

The results are contained in the `audits` sections of this object, and each of these sub-elements contains a `description` field, detailing what the metric means.

Here are, for example, some of the results, focused on performance, with their respective descriptions:

"First Meaningful Paint"

```
# Each audit point contains a description,
# that explains what this value stands for
lighthouse_report$audits$`first-meaningful-paint`$description
```

```
[1] "First Meaningful Paint measures when the primary"
[2] "content of a page is visible. [Learn"
[3] "more](https://web.dev/first-meaningful-paint)."
```

```
# We can turn the results into a data frame
lighthouse_report$audits$`first-meaningful-paint` %>%
  tibble::as_tibble()  %>%
  dplyr::select(title, score, displayValue)
```

```
# A tibble: 1 x 3
  title                 score displayValue
  <chr>                 <dbl> <chr>
1 First Meaningful Paint 0.23 5.4 s
```

"Speed Index"

```
lighthouse_report$audits$`speed-index`$description
```

```
[1] "Speed Index shows how quickly the contents of a page"
[2] "are visibly populated. [Learn"
[3] "more](https://web.dev/speed-index)."
```

```
lighthouse_report$audits$`speed-index` %>%
  tibble::as_tibble() %>%
  dplyr::select(title, score, displayValue)
```

```
# A tibble: 1 x 3
  title        score displayValue
  <chr>        <dbl> <chr>
1 Speed Index  0.56  5.4 s
```

"Estimated Input Latency"

```
lighthouse_report$audits$`estimated-input-latency`$description
```

```
[1] "Estimated Input Latency is an estimate of how long"
[2] "your app takes to respond to user input, in"
[3] "milliseconds, during the busiest 5s window of page"
[4] "load. If your latency is higher than 50 ms, users may"
[5] "perceive your app as laggy. [Learn"
[6] "more](https://web.dev/estimated-input-latency)."
```

```
lighthouse_report$audits$`estimated-input-latency` %>%
  tibble::as_tibble() %>%
  dplyr::select(title, score, displayValue)
```

```
# A tibble: 1 x 3
  title                    score displayValue
  <chr>                    <int> <chr>
1 Estimated Input Latency      1 10 ms
```

"Total Blocking Time"

```
lighthouse_report$audits$`total-blocking-time`$description
```

```
[1] "Sum of all time periods between FCP and Time to"
[2] "Interactive, when task length exceeded 50ms, expressed"
[3] "in milliseconds."
```

```
lighthouse_report$audits$`total-blocking-time` %>%
  tibble::as_tibble()  %>%
  dplyr::select(title, score, displayValue)
```

```
# A tibble: 1 x 3
  title              score displayValue
  <chr>              <int> <chr>
1 Total Blocking Time    1 30 ms
```

"Time to first Byte"

```
lighthouse_report$audits$`time-to-first-byte`$description
```

```
[1] "Time To First Byte identifies the time at which your"
[2] "server sends a response. [Learn"
[3] "more](https://web.dev/time-to-first-byte)."
```

```
lighthouse_report$audits$`time-to-first-byte` %>%
  .[c("title", "score", "displayValue")] %>%
  tibble::as_tibble()
```

```
# A tibble: 1 x 3
  title                  score displayValue
  <chr>                  <int> <chr>
1 Server response times are~    1 Root document took ~
```

Google Lighthouse also comes with a continuous integration tool, so that you can use it as a regression testing tool for your application. To know more, feel free to read the documentation[10]!

Side note on minification

Chances are that right now you are not using *minification* in your {shiny} application. Minification is the process of removing unnecessary characters from files, without changing the way the code works, to make the file size smaller. The general idea being that line breaks, spaces, and a specific set of characters are used inside scripts for human readability, and are not useful when it comes to the way a computer reads a piece of code. Why not remove them when they are served in a larger software? This is what *minification* does.

Here is an example of how minification works, taken from *Empirical Study on Effects of Script Minification and HTTP Compression for Traffic Reduction* (Sakamoto et al., 2015):

```
var sum = 0;
for ( var i = 0; i <=10; i ++ ) {
   sum += i ;
}
alert( sum ) ;
```

is minified into:

```
var sum=0;for(var i=0;i<=10;i++){sum+=i};alert(sum);
```

Both these code blocks behave the same way, but the second one will be smaller when saved to a file: this is the very core principle of minification of files. It is something pretty common to do when building web applications: on the web, every byte counts, so the smaller your external resources the better. Minification is important as the larger your resources, the longer your application will take to launch, and:

- Page launch time is crucial when it comes to ranking the pages on the web.

- The larger the resources, the longer it will take to launch the application on a mobile, notably if users are visiting your application from a 3G/4G network.

---

[10]https://github.com/GoogleChrome/lighthouse-ci/blob/master/docs/getting-started.md

And do not forget the following:

Extremely high-speed network infrastructures are becoming more and more popular in developed countries. However, we still face crowded and low-speed Wi-Fi environments on airport, cafe, international conference, etc. Especially, a network environment of mobile devices requires efficient usage of network bandwidth.

Empirical study on effects of script minification and HTTP compression for traffic reduction (Sakamoto et al., 2015)

To minify JavaScript, HTML and CSS files from R, you can use the {minifyr} (Fay, 2021g) package, which wraps the **node-minify** NodeJS library. For example, compare the size of this file from {shiny}:

```
# Displaying the file size of a CSS file from {shiny}
fs::file_size(
  system.file("www/shared/shiny.js", package = "shiny")
)
```

239K

To its minified version:

```
# Using the {minifyr} package to minify the CSS file
minified <- minifyr::minifyr_js_gcc(
  system.file("www/shared/shiny.js", package = "shiny"),
  "shinymini.js"
)
```

```
# Minifying can help you gain kilobytes
fs::file_size(minified)
```

87.5K

That might not seem like much (a couple of KB) on a small scale, but as it can be done automatically, why not leverage these small performance gains when building larger applications? Of course, minification will not suddenly make your application blazing fast, but that's something you should consider when deploying an application to production, notably if you use a lot of packages with interactive widgets: they might contain CSS and JavaScript files that are not minified.

Minification can be important notably if you expect your audience to be connecting to your app with a low bandwidth: whenever your application starts, the browser has to download the source files from the server, meaning that the larger these files, the longer it will take to render.

Note that {shiny} files are minified by default, so you will not have to re-minify them. But most packages that extend {shiny} are not, so minifying the CSS and JavaScript files from these packages might help you win some points on you Google Lighthouse report!

To do this automatically, you can add the {minifyr} commands to your deployment, be it on your CD/CI platform, or as a Dockerfile step. {minifyr} comes with a series of functions to do that:

- `minify_folder_css()`, `minify_folder_js()`, `minify_folder_html()` and `minify_folder_json()` do a bulk minification of the files found in a folder that matches the extension.
- `minify_package_js()`, `minify_package_css()`, `minify_package_html()` and `minify_package_json()` will minify the CSS and JavaScript files contained inside a package installed on the machine.

Here is what it can look like inside a `Dockerfile` (Note that you will need to install NodeJS inside the container):

```
FROM rocker/shiny-verse:3.6.3

RUN apt-get -y install curl RUN curl -sL \
   <https://deb.nodesource.com/setup_14.x> \
   | bash - RUN apt-get install -y nodejs

RUN Rscript -e 'remotes::install_github("colinfay/minifyr")'
RUN Rscript -e 'remotes::install_cran("cicerone")'
RUN Rscript -e 'library(minifyr);\
    minifyr_npm_install(TRUE);\
    minify_package_js("cicerone", minifyr_js_uglify)'
```

14.2.3   More resources about web-page performance

- Why Performance Matters - Google Web Fundamentals[11]
- Web Performance - Mozilla Web Docs[12]

---

[11]https://developers.google.com/web/fundamentals/performance/why-performance-matters
[12]https://developer.mozilla.org/en-US/docs/Web/Performance

15

15.1.1   Reactivity is awesome... until it is not

Let's face it, reactivity is awesome... until it is not. Reactivity is a common source of confusion for beginners, and a common source of bugs and bottlenecks, even for seasoned {shiny} developers. Most of the time, issues come from the fact that **there is too much reactivity**, *i.e.* we build apps where too many things happen, and some things are updated way more often than they should be, and computations are performed when they should not be, and in the end we have a hard time understanding what is really happening inside our application.

Of course, it is a nice feature to make everything react instantly to changes, but when building larger apps it is easy to create monsters, i.e. complicated, messy, reactive graphs where everything is updated too much and too often. Or worse, we generate endless reactive loops, aka "the reactive inferno" where A invalidates B which invalidates C which invalidates A which invalidates B which invalidates C, and so on.

Let's take a small example of a reactive inferno:

```
library(shiny)
library(lubridate)
ui <- function(){
  tagList(
    # Adding a first input which allow
    # to select a specific date
    dateInput(
      "date",
      "choose a date"
    ),
    # Adding a second input allowing
```

```r
    # to specify a year
    selectInput(
      "year",
      "Choose a year",
      choices = 2010:2030
    )
  )
}

server <- function(
  input,
  output,
  session
){
  # We want the year to be update whenever
  # the dateInput is updated
  observeEvent( input$date , {
    updateSelectInput(
      session,
      "year",
      selected = year(input$date)
    )
  })

  # We want the date to be update whenever
  # the selectInput is updated
  observeEvent( input$year , {
    updateDateInput(
      session,
      "date",
      value = lubridate::as_date(
        sprintf("%s-01-01", input$year)
      )
    )
  })

}

shinyApp(ui, server)
```

Here, we want to handle something pretty common:

- The user can pick a `date` and the `year` input is updated.

- And the other way round: when the `year` input changes, the `date` is updated too.

But if you try to run this in your console, it will end as a reactive inferno: date updates year that updates date that updates year, and so on.

And the more you work on your app, the more complex it gets, and the more you will be likely to end up in a reactive inferno. In this section, we will deal with reactivity, how to have more control over it, and how to share data across modules without relying on passing along reactive objects.

This application is in this state of infinite loop because it starts in a mutually inconsistent state: the `dateInput()` year value is the current year, while the `selectInput()` value is 2010. One way to solve this is to add some extra logic to the app by selecting the current year for `selectInput()`, and adding an `if` statement in the `observeEvent(input$year, {})`, as shown below.[1]

```r
library(shiny)
ui <- fluidPage(
  dateInput(
    "date",
    "choose a date"
    ),
  selectInput(
    "year",
    "Choose a year",
    choices = 2010:2030,
    # Setting a state for the year
    selected = format(
      Sys.Date(),
      "%Y"
    )
  )
)

server <- function(input, output, session) {
  observeEvent(input$date, {
    year <- format(input$date, "%Y")
    message("Changing year to ", year)
    updateSelectInput(inputId = "year", selected = year)
  })
```

---

[1]We want to thank Hadley for his help simplifying this solution https://github.com/ ThinkR-open/engineering-shiny-book/issues/276.

```
observeEvent(input$year, {
  # Preventing this update to be sent at application launch
  if (input$year != format(input$date, "%Y")) {
    date <- as.Date(ISOdate(input$year, 1, 1))
    message("Changing date to ", date)
    updateDateInput(inputId = "date", value = date)
  }
})
}

shinyApp(ui, server)
```

15.1.2 observe vs observeEvent

One of the most common features of reactive inferno is the use of `observe()`
in cases where you should use `observeEvent`. Spoiler: you should try to use
`observeEvent()` as much as possible, and avoid `observe()` as much as possible.

At first, `observe()` seems easier to implement, and feels like a shortcut as you
do not have to think about what to react to: everything gets updated without
you thinking about it. But the truth is, this stairway does not lead to heaven.

Let's stop and think about `observe()` for a minute. This function updates
every time a reactive object it contains is invalidated. Yes, this works
well if you have a small number of reactive objects in the observer, but that
gets tricky when you start adding a long list of things inside your `observe()`,
as you might be launching a computation 10 times if your reactive scope
contains 10 reactive objects that are somehow invalidated in chain. And believe
us, we have seen pieces of code where the `observe()` contains hundreds of lines
of code, with reactive objects all over the place, with one `observe()` context
being invalidated dozens of times when one input changes in the application.

For example, let's start with that:

```
## DO NOT DO GLOBAL VARIABLES, IT'S JUST TO SIMPLIFY THE EXAMPLE
# We initiate a counter that will help to track how many times
# some pieces of the code are called
i <- 0
library(shiny)
library(cli)
ui <- function(){
  tagList(
```

```r
    # We are adding a simple text input
    # that will be printed to the console
    textInput("txt", "Text")
  )
}

server <- function(input, output, session){
  observe({
    # Every time this reactive context is invalidated,
    # we add 1 to the i value
    i <<- i + 1
    # We print the i value to the console,
    # and the value of input$txt
    cat_rule(as.character(i))
    print(input$txt)
  })
}

shinyApp(ui, server)
```

Oh, and then, let's add a small `selectInput()`:

```r
i <- 0
library(shiny)
library(cli)
ui <- function(){
  tagList(
    # We are adding a simple text input
    # that will be printed to the console
    textInput("txt", "Text"),
    # We add a selectInput() to allow text transformation
    selectInput(
      "casefolding",
      "Casefolding",
      c("lower", "upper")
    )
  )
}

server <- function(input, output, session){
  observe({
    # Every time this reactive context
```

```
    # is invalidated, we add 1 to the i value
    i <<- i + 1
    # We print the i value to the console
    cat_rule(as.character(i))
    # If the user select lower, then the text is
    # passed through tolower, otherwise it's passed
    # through toupper
    if (input$casefolding == "lower") {
      print(tolower(input$txt))
    } else  {
      print(toupper(input$txt))
    }
  })
}

shinyApp(ui, server)
```

And, as time goes by, we add another control flow to our `observe()`:

```
i <- 0
library(shiny)
library(cli)
library(stringi)
ui <- function(){
  tagList(
    # We are adding a simple text input
    # that will be printed to the console
    textInput("txt", "Text"),
    # We add a selectInput() to allow text transformation
    selectInput(
      "casefolding",
      "Casefolding",
      c("lower", "upper")
    ),
    # A new checkbox to reverse (or not) the input text
    checkboxInput("rev", "reverse")
  )
}

server <- function(input, output, session){
  observe({
    # Every time this reactive context
```

```
  # is invalidated, we add 1 to the i value
  i <<- i + 1
  # We print the i value to the console
  cat_rule(as.character(i))
  # Use input_txt as a container for our input
  input_txt <- input$txt
  if (input$rev){
    # If the input$rev is select, we reverse the text
    input_txt <- stri_reverse(input_txt)
  }
  # If the user select lower, then the text is
  # passed through tolower, otherwise it's passed
  # through toupper
  if (input$casefolding == "lower") {
    print(tolower(input_txt))
  } else {
    print(toupper(input_txt))
  }
})
}

shinyApp(ui, server)
```

And it would be nice to keep the selected values in a reactive list, so that we can reuse it elsewhere. And maybe you would like to add a checkbox so that the logs are printed to the console only if checked.

```
i <- 0
library(shiny)
library(cli)
library(stringi)
ui <- function(){
  tagList(
    # We are adding a simple text input
    # that will be printed to the console
    textInput("txt", "Text"),
    # We add a selectInput() to allow text transformation
    selectInput(
      "casefolding",
      "Casefolding",
      c("lower", "upper")
    ),
```

```
    # A new checkbox to reverse (or not) the input text
    checkboxInput("rev", "reverse")
  )
}

server <- function(input, output, session){
  # We are using a reactiveValues to keep this input value
  r <- reactiveValues()
  observe({
    # Every time this reactive context
    # is invalidated, we add 1 to the i value
    i <<- i + 1
    # We print the i value to the console
    cat_rule(as.character(i))
    if (input$rev){
      # If the input$rev is select, we reverse the text
      r$input_txt <- stri_reverse(r$input_txt)
    } else {
      # Otherwise, we leave it as it is
      r$input_txt <- input$txt
    }
    # If the user select lower, then the text is
    # passed through tolower, otherwise it's passed
    # through toupper
    if (input$casefolding == "lower") {
      print(tolower(r$input_txt))
    } else  {
      print(toupper(r$input_txt))
    }
  })
}

shinyApp(ui, server)
```

Ok, now can you tell how many potential invalidation points we have here?
Three: whenever input$txt, input$rev or input$casefolding change. Of
course, three is not that much, but you get the idea.

Let's pause a minute and think about why we use observe() here. To update
the values inside r$input_txt, yes. But do we need to use observe() for, say,
updating r$input_txt under dozens of conditions, each time the user types
a letter? Possibly not.

We generally want our observer to update its content under a small, controlled
number of inputs, i.e. with a controlled number of invalidation points. And,

what we often forget is that users do not type/select correctly on the first try. No, they usually try and miss, restart, change things, amplifying the reactivity "over-happening".

Moreover, long `observe()` statements are hard to debug, and they make collaboration harder when the trigger to the observe logic can potentially live anywhere between line one and line 257 of your `observe()`. That's why (well, in 99% of cases), it is safer to go with `observeEvent`, as it allows you to see at a glance the condition under which the content is invalidated and re-evaluated. Then, if a reactive context is invalidated, **you know why**. For example, here is where the reactive invalidation can happen (lines with a ∗)[2]:

```
observe({
    i <<- i + 1
    cat_rule(as.character(i))
*   if (input$rev){
*       r$input_txt <- stri_reverse(r$input_txt)
    } else {
*       r$input_txt <- input$txt
    }
*   if (input$casefolding == "lower") {
*       print(tolower(r$input_txt))
    } else   {
*       print(toupper(r$input_txt))
    }
})
```

Whereas in this refactored code using `observeEvent()`, it is easier to identify where the invalidation can happen:

```
observeEvent( c(
*   input$rev,
*   input$txt
),{
    i <<- i + 1
    cat_rule(as.character(i))
    if (input$rev){
        r$input_txt <- stri_reverse(r$input_txt)
    } else {
```

---

[2]Of course it's an over-simplification: the reactive context will not be invalidated in all of these contexts. The idea is to illustrate how `observe()` can lead to invalidation points that are spread all across the code bloc.

```
    r$input_txt <- input$txt
  }
  if (input$casefolding == "lower") {
    print(tolower(r$input_txt))
  } else  {
    print(toupper(r$input_txt))
  }
})
```

15.1.3 Building triggers and watchers

To prevent this, one way to go is to create "flag" objects, which can be thought of as internal buttons to control what you want to invalidate: you create the button, set some places where you want these buttons to invalidate the context, and finally press these buttons.

These objects are launched with an **init** function, then these flags are triggered with **trigger()**, and wherever we want these flags to invalidate a reactive context, we **watch()** these flags.

The idea here is to get full control over the reactive flow: we only invalidate contexts when we want, making the general flow of the app more predictable. These flags are available using the {gargoyle} (Fay, 2021d) package, that can be installed from GitHub with:

```
# CRAN version
install.pacakges("gargoyle")
# Dev version
remotes::install_github("ColinFay/gargoyle")
```

- gargoyle::init("this") initiates a "this" flag: most of the time you will be generating them at the app_server() level.

- gargoyle::watch("this") sets the flag inside a reactive context, so that it will be invalidated every time you trigger("this") this flag.

- gargoyle::trigger("this") triggers the flags.

And, bonus, as these functions use the **session** object, they are available across all modules. That also means that you can easily trigger an event inside a module from another one.

This pattern is, for example, implemented in {hexmake} (Fay, 2021f) (though not with {gargoyle}), where the rendering of the image on the right is fully controlled by the "render" flag[3]. The idea here is to allow complete control over when the image is recomputed: only when trigger("render") is called does the app regenerate the image, helping us lower the reactivity of the application. That might seem like a lot of extra work, but that is definitely worth considering in the long run, as it will help in optimizing the rendering (fewer computations), and lowering the number of errors that can result from too much reactivity inside an application.

Here is a small example of this implementation, using an environment to store the value. When using this pattern, we do not rely on any reactive value invalidating the reactive context: the second result is only displayed when the "render2" flag is triggered, giving us a full control on how the reactivity is propagated.

```r
library(shiny)
library(gargoyle)
ui <- function(){
  fluidPage(
    tagList(
      # Creating an action button to launch the computation
      actionButton("compute", "Compute"),
      # Output for all runif()
      verbatimTextOutput("result"),
      # This output will change only if runif() > 0.5
      verbatimTextOutput("result2"),
      # This button will reset x$results to 0, we use it
      # to show that it won't launch a series of reactivity
      # invalidation
      actionButton("reset", "Reset x")
    )
  )
}

server <- function(
  input,
  output,
  session
){

  # Mimic an R6 class, i.e. a non-reactive object
  x <- environment()
```

---

[3]https://github.com/ColinFay/hexmake/blob/master/R/mod_right.R#L40

```r
# Creating two watchers
init("render_result", "render_result2")

observeEvent( input$compute , {
  # When the user presses compute, we launch runif()
  x$results <- runif(1)
  # Every time a new value is stored, we render result
  trigger("render_result")
  # Only render the second result if x$results is over 0.5
  if (x$results > 0.5){
    trigger("render_result2")
  }
})

output$result <- renderPrint({
  # Will be rendered every time
  watch("render_result")
  # require x$results before rendering the output
  req(x$results)
  x$results
})

output$result2 <- renderPrint({
  # This will only be rendered if trigger("render_result2")
  # is called
  watch("render_result2")
  req(x$results)
  x$results
})

observeEvent( input$reset , {
  # This resets x$results. This code block is here
  # to show that reactivity is not triggered in this app
  # unless a trigger() is called
  x$results <-  0
  print(x$results)
})

}

shinyApp(ui, server)
```

15.1.4 Using R6 as data storage

One pattern we have also been playing with is storing the app business logic inside one or more R6 objects. Why would we want to do that?

A. Sharing data across modules

Sharing an R6 object makes it simpler to create data that are shared across modules, but without the complexity generated by reactive objects, and the instability of using global variables.

Basically, the idea is to hold the whole logic of your **data reading/cleaning/processing/outputting inside an R6 class**. An object of this class is then initiated at the top level of your application, and you can pass this object to the sub-modules. Of course, this makes even more sense if you are combining it with the trigger/watch pattern from before!

```
library(shiny)
data_cleaning_ui <- function(id){
  ns <- NS(id)
  tagList(
    # Defining the UI for your first module
    # [...]
  )
}

mod_data_cleaning_server <- function(id, r6){
  moduleServer( id, function(input, output, session){
    ns <- session$ns
    observeEvent( input$launch_cleaning , {
      # Once the launch_cleaning input is triggered, we
      # use the internal method from our r6 object
      r6$clean(arg1 = input$a, arg2 = input$b)
      # Triggering the plot
      trigger("plot")
    })
  })
}

plotting_ui <- function(id){
  ns <- NS(id)
  tagList(
    # Defining the UI for your second module
    # [...]
```

```
  )
}

mod_plotting_server <- function(id, r6){
  moduleServer( id, function(input, output, session){
    ns <- session$ns
    # Rendering, inside this second module, the plot based on the
    # cleaning done in the other module
    output$plot <- renderPlot({
      # We use the trigger/watch pattern from before
      watch("plot")
      # Calling the plot() method from our R6 object
      r6$plot()
    })
  })
}


ui <- function(){
  tagList(
    # Putting our two module UIs here
    data_cleaning_ui("data_cleaning_ui"),
    plotting_ui("plotting_ui")
  )
}

server <- function(
  input,
  output,
  session
){
  # We start by creating a new instance of th
  r6 <- MyDataProcessing$new()
  # Passing this object to the two server functions
  mod_data_cleaning_server("data_cleaning_ui_1", r6)
  mod_plotting_server("plotting_ui_1", r6)

}

shinyApp(ui, server)
```

B. Be sure it is tested

During the process of building a robust {shiny} app, we strongly suggest that you test as many things as you can. This is where using an R6 for the business logic of your app makes sense: this allows you to build the whole testing of your application data logic outside of any reactive context: you simply build unit tests just as any other function.

For example, let's say we have the following R6 generator:

```r
MyData <- R6::R6Class(
  "MyData",
  # Defining our public methods, that will be
  # the dataset container, and a summary function
  public = list(
    data = NULL,
    initialize = function(data){
      self$data <- data
    },
    summarize = function(){
      summary(self$data)
    }
  )
)
```

We can then build a test for this class using {testthat}:

```r
library(testthat, warn.conflicts = FALSE)
test_that("R6 Class works", {
  # We define a new instance of this class, that will contain
  # the mtcars data.frame
  my_data <- MyData$new(mtcars)
  # We will expect my_data to have two classes:
  # "MyData" and "R6"
  expect_is(my_data, "MyData")
  expect_is(my_data, "R6")
  # And the summarize method to return a table
  expect_is(my_data$summarize(), "table")
  # We would expect the data contained in the object
  # to match the one taken as input to new()
  expect_equal(my_data$data, mtcars)
  # And the summarize method to be equal to the summary()
  #  on the input object
```

```
  expect_equal(my_data$summarize(), summary(mtcars))
})
```

```
Test passed
```

Using R6 allows to rely on these battle-tested tools when it comes to testing functions, something which is made more complex when using other patterns like reactiveValues().

15.1.5 Logging reactivity with {whereami}

Getting a good sense of how reactivity is actually working in your app is not an easy task: the reactivity logic is a graph, and it happens very quickly when you run the app, so it's very hard to follow everything.

whereami::whereami()'s (Sidi and Müller, 2019) goal is simple: informing you about where it is called, i.e. from what file and at which line, and how many times. For example, if you add the following piece of code to your app_server(), the location of the function call will be printed to the logs.

```
whereami::cat_where( whereami::whereami() )
```

```
  Running server(...) at app_server.R#9 (2)
```

Combining cat_where() will implement a reactive logging to your console while developing: that way, you can instantaneously know what reactive contexts are invalidated while using the application. Of course, you still have to implement it by hand, but that is definitely worth the effort: seeing in real time, in your console, which line is run allows you to detect unexpected behavior. For example, you will be able to see that the observeEvent() from mod_main.R#79 has been called 17 times when launching the app, which might be an unexpected behavior.

The screenshot in Figure 15.1 shows what a {whereami} log might look like, here, for the {hexmake} application.

And bonus, once the app is closed, you can get a list of all the "counters" with whereami::counter_get(), and how many times they each have been called, and plot(whereami::counter_get()) will draw a raw plot of the various counters, as shown in Figure 15.2.

```
> run_app(with_mongo = TRUE)
Loading required package: shiny

Listening on http://127.0.0.1:4923
— Running server(...) at app_server.R#4 (4) ————————————————
— Running observeEventHandler(...) at mod_manip_image.R#341 (3) ——————————
— Running observeEventHandler(...) at mod_rendering.R#74 (3) ——————
— Running renderImage(...) at mod_right.R#39 (5) ————————————
— Running renderImage(...) at mod_right.R#39 (6) ————————————
— Running renderImage(...) at mod_right.R#39 (7) ————————————
— Running renderImage(...) at mod_right.R#39 (8) ————————————
— Running renderImage(...) at mod_right.R#39 (9) ————————————
— Running observeEventHandler(...) at mod_image.R#177 (1) ——————————
— Running observeEventHandler(...) at mod_manip_image.R#592 (1) ——————
— Running observeEventHandler(...) at mod_manip_image.R#404 (1) ——————
— Running renderImage(...) at mod_right.R#39 (10) ——————————
— Running renderImage(...) at mod_manip_image.R#364 (1) ——————
— Running observeEventHandler(...) at mod_manip_image.R#404 (2) ——————
— Running renderImage(...) at mod_manip_image.R#364 (2) ——————
— Running observeEventHandler(...) at mod_manip_image.R#404 (3) ——————
— Running renderImage(...) at mod_manip_image.R#364 (3) ——————
— Running observeEventHandler(...) at mod_manip_image.R#538 (1) ——————
— Running renderImage(...) at mod_right.R#39 (11) ——————————
```

FIGURE 15.1: {whereami} output for {hexmake}.

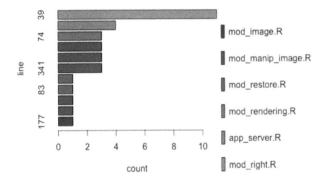

FIGURE 15.2: plot of {whereami} counters.

15.2   R does too much

15.2.1   Rendering the UI from the server side

There are many reasons we would want to change things on the UI based on what happens in the server: changing the choices of a selectInput() based on the columns of a table which is uploaded by the user, showing and hiding pieces of the app according to an environment variable, allowing the user to create an indeterminate number of inputs, etc.

Chances are that to do that, you have been using the `uiOutput()` and `renderUI()` functions from {shiny} (Chang et al., 2021a). Even if convenient, and the functions of choice in some specific context, this pair of functions makes R do a little bit too much: you are making R regenerate the whole UI component instead of changing only what you need, which can be a suboptimal, be it from the user point of view, or from a developer perspective.

One of the instance in which this pattern might not be optimal is in the case where your visitors do not have an high-speed internet or when visiting and using a smartphone, contexts where every byte counts. Rendering large elements from the server side in your {shiny} app means that these elements will have to transit through the socket, i.e. they need to be sent by the server, and downloaded by the browser. In this case, the smaller the message size the better!

From the developer perspective, you will create code that is harder to reason about, as we are used to having the UI parts in the UI functions (but that is not related to performance).

Here are three strategies to code without `uiOutput()` and `renderUI()`.

A. Implement UI events in JavaScript

Mixing languages is better than writing everything in one, if and only if using only that one is likely to overcomplicate the program.

The Art of UNIX Programming (Raymond, 2003)

We will see in the last chapter of this book how you can integrate JS inside your {shiny} app, and how even basic functions can be useful for making your app server smaller. For example, compare:

```
library(shiny)
ui <- function(){
  tagList(
    # Adding a button with an onclick event,
    # that will show or hide the plot
    actionButton(
```

```
      "change",
      "show/hide graph",
      # The toggle() function hide or show the queried element
      onclick = "$('#plot').toggle()"
    ),
    plotOutput("plot")
  )
}

server <- function(
  input,
  output,
  session
){
  output$plot <- renderPlot({
    # This renderPlot will only be called once
    cli::cat_rule("Rendering plot")
    plot(iris)
  })
}

shinyApp(ui, server)
```

to

```
library(shiny)
ui <- function(){
  tagList(
    # We use a pattern without JavaScript
    actionButton("change", "show/hide graph"),
    plotOutput("plot")
  )
}

server <- function(
  input,
  output,
  session
){

  output$plot <- renderPlot({
    # Here, every time the button is clicked, this reactive
```

```
    # context will be invalidated, and the code re-evaluated
    cli::cat_rule("Rendering plot")
    # Simulate a show and hide pattern
    req(input$change %% 2 == 0)
    plot(iris)
  })

}

shinyApp(ui, server)
```

The result is the same, but the first version is shorter and easier to understand: we have one button, and the behavior of the button is self-contained. The second solution redraws the plot every time the `reactiveValues` is updated, making R compute way more than it should, whereas with the JavaScript-only solution, the plot is not recomputed every time you need to show it: the plot is drawn by R only once.

At a local level, the improvements described in this section will not make your application way faster: for example, rendering UI elements (let's say rendering a simple title) will not be computationally heavy. But at a global level, less UI computation from the server side helps the general rendering of the app: let's say you have an output that takes 3 seconds to run, then if the whole UI + output is to be rendered on the server side, the whole UI stays blank until everything is computed.

Compare:

```
library(shiny)
ui <- function(){
  tagList(
    # We make the whole UI be generated by R
    uiOutput("caption")
  )
}

server <- function(
  input,
  output,
  session
){
  output$caption <- renderUI({
    # Simulate something that takes 3 seconds to run
```

```
    Sys.sleep(3)
    # Returning the UI
    tagList(
      h3("test"),
      shinipsum::random_text(10)
    )

  })
}

shinyApp(ui, server)
```

to

```
library(shiny)
ui <- function(){
  tagList(
    # Only the text input will be rendered by R
    h3("test"),
    textOutput("caption")
  )
}

server <- function(
  input,
  output,
  session
){
  output$caption <- renderText({
    # Here, we only render the text, not the whole UI
    Sys.sleep(3)
    shinipsum::random_text(10)
  })
}

shinyApp(ui, server)
```

In the first example, the UI will wait for the server to have rendered, while in the second we will first see the title, then the rendered text after a few seconds. That approach makes the user experience better: they know that something is happening, while a completely blank page is confusing.

Also, because R is single threaded, manipulating DOM elements from the server side causes R to be busy doing these DOM manipulations while it could be computing something else. And let's imagine it takes a quarter of a second to render the DOM element. That is a full second for rendering four of them, while R should be busy doing something else!

B. update* inputs

Almost every {shiny} input, even the custom ones from packages, come with an `update_` function that allows us to change the input values from the server side, instead of re-creating the UI entirely. For example, here is a way to update the content of a `selectInput` from the server side:

```r
library(shiny)
ui <- function(){
  tagList(
    # We start the selectInput empty
    selectInput("species", "Species", choices = NULL),
    # The selectInput will be populate
    # when the update button is pressed
    actionButton("update", "Update")
  )
}

server <- function(
  input,
  output,
  session
){
  observeEvent( input$update , {
    # Update the selectInput with the species from iris
    spc <- unique(iris$Species)
    updateSelectInput(
      session,
      "species",
      choices = spc,
      selected = spc[1]
    )
  })

}

shinyApp(ui, server)
```

This switch to `updateSelectInput` makes the code easier to reason about as the `selectInput` is where it should be: inside the UI, instead of another pattern where we would use `renderUI()` and `uiOutput()`. Plus, with the `update` method, we are only changing what is needed, not re-generating the whole input.

C. insertUI and removeUI

Another way to dynamically change what is in the UI is with `insertUI()` and `removeUI()`. It is more global than the solution we have seen before with setting the `reactiveValue` to `NULL` or to a value, as it allows us to target a larger UI element: we can insert or remove the whole input, instead of having the DOM element inserted but empty. This method allows us to have a smaller DOM: `<div>` that are not rendered are not generated empty, they are simply not there.

Two things to note concerning this method, though:

- Removing an element from the app will not delete the input from the input list. In other words, if you have `selectInput("x", "x")`, and you remove this input using `removeUI()`, you will still have `input$x` in the server.

For example, in the following example, the `input$val` value will not be removed once you have called `removeUI(selector = "#val")`.

```r
library(shiny)
ui <- function(){
  tagList(
    # Creating a text input that will be removed
    #  from the UI whenever the remove button is pressed
    textInput("value", "Value", "place"),
    actionButton("remove", "Remove UI")
  )
}

server <- function(
  input,
  output,
  session
){

  observeEvent( input$remove , {
    # When the button is pressed,
    #  the textInput will be removed from the UI
```

```
   removeUI(selector = "#value")
})

observe({
  # We observe input$value every second.
  # You'll realize that even after the UI
  # is removed, input$value is still available.
  invalidateLater(1000)
  print(input$value)
})

}

shinyApp(ui, server)
```

- Both these functions take a `jQuery` selector to select the element in the UI. We will introduce these selectors in Chapter 17.

15.2.2 Too much data in memory

If you are building a {shiny} application, there is a great chance you are building it to analyze data. If you are dealing with large datasets, **you should consider deporting the data handling and computation to an external database system: for example, to an SQL database**. Why? Because these systems have been created to handle and manipulate data on disk: in other words, it will allow you to perform operations on your data without having to clutter R memory with a large dataset.

For example, if you have a `selectInput()` that is used to perform a filter on a dataset, you can do that filter straight inside SQL, instead of bringing all the data to R and then doing the filter. That is even more necessary if you are building the app for a large number of users: for example if one {shiny} session takes up to 300MB, multiply that by the number of users that will need one session, and you will have a rough estimate of how much RAM you will need. On the contrary, if you reduce the data manipulation so that it is done by the back-end, you will have, let's say, one database with 300MB of data, so the database size will remain (more or less constant), and the only RAM used by {shiny} will be the data manipulation, not the data storage. That's even more true now that almost any operation you can do today in {dplyr} (Wickham et al., 2021b) would be doable with an SQL back-end, and that is the purpose of the {dbplyr} (Wickham et al., 2021c) package: translates {dplyr} code into SQL.

If using a database as a back-end seems a little bit far-fetched right now, that is how it is done in most programming languages: if you are building a web app with NodeJS or Python for example, and need to interact with data, nothing will be stored in RAM: you will be relying on an external database to store your data. Then your application will be used to make queries to this database back-end.

15.3 Reading data

{shiny} applications are a tool of choice when it comes to analyzing data. But that also means that these data have to be imported/read at some point in time, and reading data can be time consuming. How can we optimize that? In this section, we will take a look at three strategies: including datasets inside your application, using R packages for fast data reading, and when and why you should move to an external database system.

15.3.1 Including data in your application

If you are building your application using the {golem} (Fay et al., 2021) framework, you are building your application as a package. R packages provide a way to include internal datasets, which can then be used as objects inside your app. This is the solution you should go for if your data are never to rarely updated: the datasets are created during package development, then included inside the build of your package. The plus side of this approach is that it makes the data fast to read, as they are serialized as R native objects.

To include data inside your application, you can use the `usethis::use_data_raw( name = "my_dataset", open = FALSE )` command, which is inside the `02_dev.R` script inside the `dev/` folder of your source application (if you are building the app with {golem}). This will create a folder called `data-raw` at the root of your application folder, with a script to prepare your dataset. Here, you can read the data, modify it if necessary, and then save it with `usethis::use_data(my_dataset)`. Once this is done, you will have access to the `my_dataset` object inside your application.

This is, for example, what is done in the {tidytuesday201942} (Fay, 2021k) application, in data-raw/big_epa_cars.R[4]: the CSV data are read there, and then used as an internal dataset inside the application.

---

[4]https://github.com/ColinFay/tidytuesday201942/blob/master/data-raw/big_epa_cars.R

15.3.2   Reading external datasets

Other applications use data that are not available at build time: they are created to analyze data that are uploaded by users, or maybe they are fetched from an external service while using the app (for example, by calling an API). When you are building an application for the "user data" use case, the first thing you will need is to provide users a way to upload their dataset: `shiny::fileInput()`.

One crucial thing to keep in mind when it comes to using user-uploaded files is that you have to be (very) strict with the way you handle files:

- Always specify what type of file you want: `shiny::fileInput()` has an `accept` parameter that allows you to set one or more MIME types[5] or extensions. When using this argument (for example, with `text/csv`, `.csv`, or `.xslx`), the user will only be able to select a subset of files from their computer: the ones that match the type.
- Always perform checks once the file is uploaded, even more if it is tabular data: column type, naming, empty rows, etc. The more you check the file for potential errors, the less your application is likely to fail to analyze this uploaded dataset.
- If the data reading takes a while, do not forget to add a visual progression cue: a `shiny::withProgress()` or tools from the {waiter}[6] package.

Whenever you offer a user the possibility to upload anything, you can be sure that at some point, they will upload a file that will make the app crash. By setting a specific MIME type and by doing a series of checks once the file is uploaded, you will make your application more stable. Finally, having a visual cue that "something is happening" is very important for the user experience, because "something is happening" is better than not knowing what is happening, and it may also prevent the user from clicking again and again on the upload button, or worse, they will stop using the app.

Now that we have our `fileInput()` set, how do we read these data as fast as possible? There are several options depending on the type of data you are reading. Here are some packages that can make the file reading faster:

- For a tabular, flat dataset (typically csv, tsv, or text), {vroom} (Hester and Wickham, 2021) can read data at a 1.40 GB/sec/sec speed. The `fread()` function from {data.table} (Dowle and Srinivasan, 2021) is also fast at reading delimited files.
- For JSON files, {jsonlite} (Ooms, 2014). Or more recently, {RcppSimdJSON} (Eddelbuettel and Knapp, 2021), which is a binding to the `simdjson` C++ library.

---

[5]https://en.wikipedia.org/wiki/Media_type
[6]https://github.com/JohnCoene/waiter

- If you need to read Excel files inside your app, {readxl} (Wickham and Bryan, 2019) offers a binding to the RapidXML[7] C++ library, which reads Excel files fast.
- Most files exported from statistical software (SAS, SPSS, etc.) can be read using either the {foreign} (R Core Team, 2020) or {haven} (Wickham and Miller, 2021) packages.

15.3.3 Using external databases

Another type of data analyzed in a shiny application is data that is contained inside an external database. Databases are heavily used in the data science world and in software engineering as a whole. Databases come with APIs and drivers that help retrieve and transfer data: be it SQL, NoSQL, or even a graph.

Using a database is one of the solutions for making your app smaller and more efficient in the long run, especially if you need to scale your app to thousands of visitors. Indeed, **if you plan on having your app scale to numerous people, that will mean that a lot of R processes will be triggered. And if your data is contained in your app, this will mean that each R process will take a significant amount of RAM if the dataset is large.** For example, if your dataset alone takes ~300 MB of RAM, that means that if you want to launch the app 10 times, you will need ~3GB of RAM. On the other hand, if you decide to switch these data to an external database, it will lower the global RAM need: the DB will take these 300MB of data, and each shiny application will make a request to the database. For instance, if the database needs 300MB, and one shiny app 50MB, then 10 apps will be 300MB (for the DB) + 50MB * 10 (for the 10 apps). In practice, other things are to be considered: making database requests can be computationally expensive, and might need some network adjustments, but you get the idea.

How does one choose between database back-end? Well, first of all you need to see what is available in the environment the application will be deployed: maybe the company you are building the application for already has database servers deployed. If ever you are free to choose any database as a back-end, your choice should be driven by what kind of operations you want to make on these databases. **For example, SQL databases are designed to store tabular data, and they tend to be very fast when it comes to reading data: so if you have one or more large data.frames you want to use inside your application, and with no specific update of these data, an SQL back-end can be the perfect choice.** On the other hand, a NoSQL database like MongoDB will be faster when it comes to doing write operations, and can store any kind of object: for example, {hexmake} can use

---

[7]http://rapidxml.sourceforge.net/

a MongoDB back-end to store RDS files. But that comes with a price: read calls are a little bit slower, and you might have to work a little bit more on handling the JSON results that come out of MongoDB. Another example of an app that uses on an external database is {databasedemo}, available at engineering-shiny.org/databasedemo/[8]. Feel free to follow this link for more information about this application!

Covering all the available types of databases and the packages associated with each is a very, very large topic: there are dozens of database systems, and as many (if not more) packages to interact with them. For more extensive coverage of using databases in R, please follow these resources:

- Databases using R[9], the official RStudio documentation around databases and R.

- colinfay/r-db[10], a Docker image that bundles the toolchain for a lot of database systems for R.

- CRAN Task View: Databases with R[11]: the official task view from CRAN with a series of packages for database manipulation

15.3.4   Data-source checklist

How to choose between these three methodologies:

Choice	Update	Size
Package data	Never to very rare	Low to medium
Reading files	Uploaded by Users	Preferably low
External DataBase	Never to Streaming	Low to Big

---

[8]https://engineering-shiny.org/databasedemo/
[9]https://db.rstudio.com/
[10]https://colinfay.me/r-db/
[11]https://cran.r-project.org/web/views/Databases.html

16

Optimizing {shiny} Code

16.1 Optimizing R code

In its core, {shiny} runs R code on the server side. To be efficient, the R code computing your values and returning results also has to be optimized.

Optimizing R code is a very broad topic, and it would be possible to write a full book about it. In fact, a lot of books and blog posts already cover this topic. Instead of re-writing these books, we will try to point to some crucial resources you can refer to if you want to get started optimizing your R code.

- Efficient R programming (Gillespie and Lovelace, 2017), has a series of methods you can quickly put into practice for more efficient R code.

- Advanced R (Wickham, 2019) has a chapter about optimizing R code (number 24). In the rest of this chapter, we will be focusing on how to optimize {shiny} specifically.

16.2 Caching elements

16.2.1 What is caching?

Caching is the process of storing resources-intensive results so that when they are needed again, your program can reuse the result another time without having to redo the computation again. This is particularly useful for computation that will always return the same result, and should never be used if you expect the result could vary from one function call to the other.

How does it work? Let's make a brief parallel with the human brain, and imagine that you know that you will need to use a phone number many times during the day, and for the purpose of this thought experiment, you are completely unable to remember it.[1] What are you going to do? There are two solutions here: either you look in the phone book or in your phone contact list every time you need it, which takes a couple of seconds every time, or you use a post-it that you put on your computer screen with the number on it, so that you have direct access to it when you need it. It takes a couple of seconds the first time you look for the number, but it is almost instantaneous the next times you need it.

This is what caching does: **it stores the result of an expensive computation, so that the next time you need the very same information again, you can read the result instead of redoing the full computation.** The downside is that you only have limited space on your screen: when your screen is covered by sticky notes, you cannot store any more notes.[2]

In the context of an interactive application built with {shiny}, it makes sense to cache data structures: users tend to repeat what they do, or go back and forth between parameters. For example, if you have a graph which is taking 2 seconds to render (which is quite common in {shiny}, notably when relying on {ggplot2} (Wickham et al., 2021a)), you do not want these 2 seconds to be repeated over and over again when users switch from one parameter to another. In that case, it does make sense to cache the result: if you call `ploting_function(input$selection)` twice with the same value for `input$selection`, and you are sure that this plot will be the same every time, you can cache it. In other words, instead of recomputing the graph on each `input$selection` change, you can cache the plot the first time it is generated, and then the application will read the cache instead of re-doing the computation.

Same goes for queries to a database: if a query is done with the same parameters, and you know that they will return the same result, there is no need to ask the database again and again—ask the cache to retrieve the data.

Keep in mind that this caching mechanism is only to be used when **the data don't change**. For example, if you are calling a database which is updated on a regular basis, you might not want to cache the results of a function. In that specific case, you will want the query to be performed every time the function is called, so that you get fresh data.

---

[1] Anyway, now that we all have smartphones, who still remembers phone numbers?

[2] In that case, you can either hide pre-existing sticky notes, or buy a bigger screen. But we are not here to talk about cache management theory. If you are interested in reading more about caching theory, we suggest the excellent *Algorithms to Live By*, by Brian Christian and Tom Griffiths (Christian and Griffiths, 2016).

16.2.2 Native caching in R

At least two packages in R implement caching of functions (also called memoization): {R.cache} (Bengtsson, 2021), and {memoise} (Wickham et al.,
2021e). They both more or less work the same way: you will call a memoization function on another function, and cache is created for this function
output, based on the arguments value. Then every time you call this function
again with the same parameters, the cache is returned instead of computing
the function another time. For example, if computing your data once takes
5 seconds with the parameter n = 50, the next time you will be calling this
function with n = 50, instead of recomputing, R will go and fetch the value
stored in cache.

Here is a simple example with {memoise}:

```r
library(memoise)
library(tictoc)
# We define a function that sleeps for a given number of seconds,
# then return the time
sleep_and_return_time <- function(seconds = 1){
  Sys.sleep(seconds)
  return(Sys.time())
}
# "Memoising" this function
msleep_and_return_time <- memoise(sleep_and_return_time)
# We use the {tictoc} package to count the time to run the code
tic()
# This will sleeep for 2 seconds and return the time
msleep_and_return_time(2)
```

```
[1] "2021-07-16 14:40:53 CEST"
```

```r
# The code should have taken around 2 seconds to run
toc()
```

```
2.008 sec elapsed
```

```
# We launch a new recording
tic()
# This memoised function will return immediately,
# without sleeping
msleep_and_return_time(2)
```

```
[1] "2021-07-16 14:40:53 CEST"
```

```
toc()
```

```
0.006 sec elapsed
```

Let's try with another example that might look more like what we can find in a {shiny} app: connecting to a database, using the {DBI} (R Special Interest Group on Databases (R-SIG-DB) et al., 2021) and {RSQLite} (Müller et al., 2021) packages:

```
# We create an in-memory database using SQLite
con <- DBI::dbConnect(
  RSQLite::SQLite(),
  dbname = ":memory:"
)
```

```
# Writing a large dataset to the db
DBI::dbWriteTable(
  con,
  "diams",
  # This table will have 539400 rows
  dplyr::bind_rows(
    purrr::rerun(10, ggplot2::diamonds)
  )
)
```

```
# We memoise the dbGetQuery,
#  so that every time this function is called with
# the same parameters,
# the SQL query is not actually run,
#  but the results are fetched from the cache
m_get_query <- memoise(DBI::dbGetQuery)
```

```
# We call a function the first time,
# with the connection object and an SQL query
tic()
res_a <- m_get_query(
  con,
  "SELECT * FROM diams WHERE cut = 'Ideal'"
)
toc()
```

1.251 sec elapsed

```
# We call this function a second time,
# with the same parameters
tic()
res_b <- m_get_query(
  con,
  "SELECT * FROM diams WHERE cut = 'Ideal'"
)
toc()
```

0.005 sec elapsed

```
# Let's check that the two are equal
setequal(res_a, res_b)
```

[1] TRUE

```
# We now try with a new SQL code (cut = 'Good')
tic()
res_c <- m_get_query(
  con,
  "SELECT * FROM diams WHERE cut = 'Good'"
)
toc()
```

0.384 sec elapsed

```
# The function has effectively returned a different result
setequal(res_a, res_c)
```

```
[1] FALSE
```

Note that you can change where the cache is stored by {memoise}. Here, we will save it in a random directory (do not do this in production).

```
random_dir <- fs::path(
  paste(
    sample(
      letters,
      10
    ),
    collapse = ""
  )
)
random_dir
```

```
xawcubtjzp
```

```
# We create a directory in the current working directory
fs::dir_create(random_dir)
# We use this directory as the cache_filesystem for {memoise}
local_cache_folder <- cache_filesystem(random_dir)
# The memoised function will use this directory for cache
m_get_query <- memoise(
  DBI::dbGetQuery,
  cache = local_cache_folder
)
# Run the function twice
res_a <- m_get_query(
  con,
  "SELECT * FROM diams WHERE cut = 'Ideal'"
)
res_b <- m_get_query(
  con,
  "SELECT * FROM diams WHERE cut = 'Good'"
)
```

```
res_c <- m_get_query(
  con,
  "SELECT * FROM diams WHERE cut = 'Good'"
)
# The random directory now contains two objects,
# one for each memoized call
fs::dir_tree(random_dir)
```

```
xawcubtjzp
+-- 3764a0a4950cb30b
\-- c295e060ea7d77c1
```

As you can see, we now have two cache objects inside the directory we have
specified as a `cache_filesystem`.

16.2.3 Caching in {shiny}

We can apply what we have just seen with {memoise}, for example, to render
a table:

```
library(memoise)
# We create an in-memory database using SQLite
con <- DBI::dbConnect(
  RSQLite::SQLite(),
  dbname = ":memory:"
)

# Writing a large dataset to the db
DBI::dbWriteTable(
  con,
  "diams",
  # This table will have 539400 rows
  dplyr::bind_rows(
    purrr::rerun(10, ggplot2::diamonds)
  )
)

fct_sql <- function(cut, con){
  # NEVER EVER SPRINTF AN SQL CODE LIKE THAT
  # IT'S SENSITIVE TO SQL INJECTIONS, WE'RE
```

```r
  # DOING IT FOR THE EXAMPLE
  cli::cat_rule("Calling the SQL db")
  results <- DBI::dbGetQuery(
    con, sprintf(
      "SELECT * FROM diams WHERE cut = '%s'",
      cut
    )
  )
  head(results)
}

# Using a local cache
cache_dir <- cache_filesystem("cache")
memoised_fct_sql <- memoise(fct_sql, cache = cache_dir)
```

Then, it can be used in an app:

```r
library(shiny)
ui <- function(){
  tagList(
    # The user can select one of the cut from ggplot2::diamonds,
    # {shiny} will then query the SQL database to retrieve the
    # first rows of the result
    selectInput("cut", "cut", unique(ggplot2::diamonds$cut)),
    tableOutput("tbl")
  )
}

server <- function(
  input,
  output,
  session
){

  # Rendering the table of the SQL call
  output$tbl <- renderTable({
    # Using a memoised function allows to prevent from
    # calling the SQL database every time the user inputs
    # a change
    memoised_fct_sql(input$cut, con)
  })
```

```
}
```

```
shinyApp(ui, server)
```

You will see that the first time you run this piece of code, it will take a couple of seconds to render the table for a new `input$cut` value. But if you re-select this input a second time, the output will show instantaneously.

Since version `1.6.0`, {shiny} (Chang et al., 2021a) has two caching functions: `renderCachedPlot()` and `bindCache()` (note that `renderCachedPlot()` is in {shiny} since version `1.2.0`).

`renderCachedPlot()` behaves more or less like the `renderPlot()` function, except that it is tailored for caching. The extra arguments you will find are `cacheKeyExpr` and `sizePolicy`: the former is the list of inputs and values that allow you to cache the plot—every time these values and inputs are the same, they produce the same graph, so {shiny} will be fetching inside the cache instead of computing the value another time. `sizePolicy` is a function that returns a `width` and a `height`, which are used to round the plot dimension in pixels, so that not every pixel combination is generated in the cache.

The good news is that converting existing `renderPlot()` functions to `renderCachedPlot()` is pretty straightforward in most cases: take your current `renderPlot()`, and add the cache keys.[3]

Here is an example:

```
library(shiny)
ui <- function(){
  tagList(
    # We select a data.frame to plot
    selectInput(
      "tbl",
      "Table",
      c("iris", "mtcars", "airquality")
    ),
    # This plotOutput will be cached
    plotOutput("plot")
  )
}
```

---

[3]In some cases you will have to configure the size policy, but in most cases the default values work just well.

```r
server <- function(
  input,
  output,
  session
){

  # The cache mechanism is made available by renderCachedPlot
  output$plot <- renderCachedPlot({
    # Plotting the selected data.frame
    plot( get(input$tbl) )
  }, cacheKeyExpr = {
    # List here all the reactive expression that will
    # be used as cache key when running the app,
    # you will see that the first time you plot one
    # graph, it takes a couple of seconds,
    # but the second time, it's almost
    # instantaneous
    input$tbl
  })

}

shinyApp(ui, server)
```

If you try this app, the first rendering of the three plots will take a little bit of time, but every subsequent rendering of the plot is almost instantaneous.

bindCache(), a new function from version 1.6.0, offers a more general approach, as it can cache any reactive expression.

```r
library(shiny)
ui <- function(){
  tagList(
    # Select a number of row to sample from mtcars
    sliderInput(
      "nrows",
      "Number of rows",
      1,
      nrow(mtcars),
      10
    ),
    tableOutput("tbl")
  )
```

```
}

server <- function(
  input,
  output,
  session
){

  # The random sample will always be the same
  # Whenever input$nrows is the same
  output$tbl <- renderTable({
    dplyr::sample_n(mtcars, input$nrows)
  }) %>%
    bindCache({
      input$nrows
    })
}

shinyApp(ui, server)
```

Caching is a nice way to make your app faster, even more if you expect your output to be stable over time: if the plot created by a series of inputs stays the same throughout your app lifecycle, it is worth thinking about implementing on-disk caching. With {memoise}, you can also use remote caching, in the form of Amazon S3 storage or with Google Cloud Storage. See also the {bank}[4] package for database caching of {shiny} expressions.

If your application needs "fresh" data every time it is used, for example because data in the SQL database are updated every hour, cache will not be of much help here. On the contrary, the same function inputs will render different output depending on when they are called.

One other thing to remember is that, just like our computer screen from our phone number example from before, you do not have unlimited space when it comes to cache storage: storing a large amount of cache will take space on your disk.

For example, from our stored cache from before:

```
dir_i <- fs::dir_info("cache")[, "size", drop = FALSE]
head(dir_i)
```

---

[4]https://github.com/ThinkR-open/bank

```
# A tibble: 6 x 1
        size
  <fs::bytes>
1         954
2         440
3         922
4         928
5         406
6         939
```

Managing cache at a system level is a very vast, fascinating topic that we
cannot cover here, but note that the most commonly accepted rule for deleting
cache is called **LRU**, for **Least Recently Used**. The underlying principle
of this approach is that users tend to need what they have needed recently:
hence the more a piece of data has been used recently, the more likely it is
that it will be needed soon. And this can be retrieved with:

```
dir_at <- fs::dir_info("cache")[, "access_time", drop = FALSE]
head(dir_at)
```

```
# A tibble: 6 x 1
  access_time
  <dttm>
1 2020-11-19 21:01:04
2 2020-10-05 08:59:56
3 2021-07-16 12:37:27
4 2021-07-16 12:39:34
5 2020-10-05 08:59:56
6 2020-11-19 21:01:04
```

Hence, when using cache, it might be interesting to periodically remove the
oldest used cache, so that you can regain some space on the server running
the application.

16.3   Asynchronous in {shiny}

One of the drawbacks of {shiny} is that as it is running on top of R, it is
single threaded, meaning that each computation is run in sequence, one after
the other. Well, at least natively, as methods have emerged to run pieces of
code in parallel.

16.3.1 How to

To launch code blocks in parallel, we will use a combination of two packages, {future} (Bengtsson, 2020a) and {promises} (Cheng, 2021), and a reactiveValue(). {future} is an R package whose main purpose is to allow users to send code to be run elsewhere, i.e. in another session, thread, or even on another machine. {promises}, on the other hand, is a package providing structure for handling asynchronous programming in R.[5]

A. Asynchronous for cross-sessions availability

The first type of asynchronous programming in {shiny} **allows non-blocking programming in a cross-session context**. In other words, it is a programming method which is useful in the context of running one {shiny} session that is accessed by multiple users. Natively, in {shiny}, if *user1* launches a 15-seconds computation, then *user2* has to wait for this computation to finish before launching their own 15-seconds computation, and *user3* has to wait the 15 seconds of *user1* plus the 15 seconds for user, etc.

With {future} and {promises}, each long computation is sent to be run somewhere else, so when *user1* launches their 15-seconds computation, they are not blocking the R process for *user2* and *user3*.

How does it work?[6] {promises} comes with two operators which will be useful in our case, %...>% and %...!%: the first being "what happens when the future() is solved?" (i.e. when the computation from the future() is completed), and the second is "what happens if the future() fails?" (i.e. what to do when the future() returns an error).

Here is an example of using this skeleton:

```
library(future)
library(promises)
# We're opening several R session (future specific)
plan(multisession)
# We send our code to be run in another session
future({
  Sys.sleep(3)
  return(rnorm(5))
```

---

[5]If you are familiar with promises in JavaScript, {promises} is an implementation of this structure into R.

[6]We're providing a short introduction with key concepts, but for a more thorough introduction, please refer to the online documentation[7].

```
}) %...>% (
  # When the code is returned, we print the result
  function(result){
    print(result)
  }
) %...!% (
  # If ever the code from the future() returns an error,
  # we throw an error to the console
  function(error){
    stop(error)
  }
)
```

If you run this in your console, you will see that you have access to the R
console directly after launching the code. And a couple of seconds later (a
little bit more than 3), the result of the `rnorm(5)` will be printed to the
console.

Note that you can also write a one-line function with . as a parameter, instead
of building the full anonymous function (we will use this notation in the rest
of the chapter):

```
library(future)
library(promises)
plan(multisession)
# Same code as before, using the anonymous notation
future({
  Sys.sleep(15)
  return(rnorm(5))
}) %...>%
  print(.) %...!%
  stop(.)
```

Let's port this to {shiny}:

```
library(shiny)
library(future)
library(promises)
plan(multisession)
ui <- function(){
  tagList(
```

```r
    # This will receive the output of the future
    verbatimTextOutput("rnorm")
  )
}

server <- function(
  input,
  output,
  session
){
  output$rnorm <- renderPrint({
    # Sending the rnorm to be run in another session
    future({
      Sys.sleep(3)
      return(rnorm(5))
    }) %...>%
      print(.) %...!%
      stop(.)
  })
}

shinyApp(ui, server)
```

If you have run this, it does not seem like a revolution: but trust us, the
Sys.sleep() is not blocking as it allows other users to launch the same computation at the same moment.

B. Inner-session asynchronicity

In the previous section, we implemented cross-session asynchronicity, meaning
that the code is non-blocking, but **when two or more users access the
same app: the code is still blocking at an inner-session level**. In other
words, the code in the **renderPrint()** will still block the rest of the app for
a single user.

Let's have a look at this code:

```r
library(shiny)
ui <- function(){
  tagList(
    # This will receive the output of the future
    verbatimTextOutput("rnorm"),
```

```r
    # This plot will only be drawn when the future
    # is resolved
    plotOutput("plot")
  )
}

server <- function(
  input,
  output,
  session
){
  output$rnorm <- renderPrint({
    # Sending the rnorm to be run in another session
    # At this point, {shiny} is waiting for the future
    # to be solved before doing anything else
    future({
      Sys.sleep(3)
      return(rnorm(5))
    }) %...>%
      print(.) %...!%
      stop(.)
  })

  # This plot will only be drawn once the future is resolved
  output$plot <- renderPlot({
    plot(iris)
  })
}

shinyApp(ui, server)
```

Here, you would expect the plot to be available before the rnorm(), but it is
not: {promises} is still blocking at an inner-session level, so elements are still
rendered sequentially. To bypass that, we will need to use a reactiveValue()
structure.

```r
library(shiny)
library(promises)
library(future)
plan(multisession)

ui <- function(){
```

```
  tagList(
    # This will receive the output of the future
    verbatimTextOutput("rnorm"),
    # This plot will be drawn before the future is resolved
    plotOutput("plot")
  )
}
server <- function(
  input,
  output,
  session
) {

  # Initiating a reactiveValues that will receive the
  # results from the future
  rv <- reactiveValues(
    output = NULL
  )

  future({
    Sys.sleep(5)
    rnorm(5)
  }) %...>%
    # When the future is resolved, we assign the
    # output to rv$output
    (function(result){
      rv$output <- result
    }) %...!%
    # If ever the future outputs an error, we switch
    # back to NULL for rv$output, and throw a warning
    # with the error
    (function(error){
      rv$output <- NULL
      warning(error)
    })

  # output$rnorm will be printed whenever rv$output
  # is available (i.e. after around 5 seconds)
  output$rnorm <- renderPrint({
    req(rv$output)
  })

  # output$plot will be drawn immediately
  output$plot <- renderPlot({
```

```
    plot(iris)
  })
}

shinyApp(ui, server)
```

Let's detail this code step-by-step:

- `rv <- reactiveValues` creates a `reactiveValue()` that will contain `NULL`, and which will serve the content of `renderPrint()` when the `future()` is resolved. It is initiated as `NULL` so that the `renderPrint()` is silent at launch.

- `%...>% rv$output <- result %...!%` is the {promises} structure we have seen before.

- `%...!% (function(error){ rv$output <- NULL ; warning(e) })` is what happens when the `future({})` fails: we are setting the `rv$res` value back to `NULL` so that the `renderPrint()` does not fail and prints an error in case of failure.

C. Potential pitfalls of asynchronous {shiny}

There is one thing to be aware of if you plan on using this async methodology: you are not in a sequential context anymore. Hence, the first `future({})` you will send is not necessarily the first you will get back. For example, if you send SQL requests to be run asynchronously and each call takes between 1 and 10 seconds to return, there is a chance that the first request to return will be the last one you sent. To handle that, we can adopt two different strategies, depending on what we need:

- We need only the last expression sent. In other words, if we send three expressions to be evaluated somewhere, we only need to get back the last one. To handle that, the best way is to have an id that is also sent to the future, and when the future comes back, we check that this id is the one we are expecting. If it is, we update the `reactiveValues()`. If it is not, we ignore it.

```
library(shiny)
library(promises)
library(future)
plan(multisession)
```

```r
ui <- function(){
  tagList(
    # This button trigger a future, we can click several times
    # on it when the app is running
    actionButton("go", "go"),
    # This will receive the output of the future
    verbatimTextOutput("rnorm"),
    # This plot will be drawn before the future is resolved
    plotOutput("plot")
  )
}

server <- function(
  input,
  output,
  session
) {

  # In our reactiveValues, we also keep track
  # of the latest sent id
  rv <- reactiveValues(
    res = NULL,
    last_id = 0
  )


  observeEvent( input$go , {
    # When the user clicks on the button, the last_id
    # is incremented of one
    rv$last_id <- rv$last_id + 1
    last_id <- rv$last_id

    # We send the code to be run in the future. One out of
    # two calls will sleep for 3 seconds
    future({
      if (last_id %% 2 == 0){
        Sys.sleep(3)
      }
      # We return from the future the id of the current
      # code block
      list(
        id = last_id,
        res = rnorm(5)
```

```
    )
  }) %...>%
    (function(result){
      # Printing to the console which future
      # we are coming from
      cli::cat_rule(
        sprintf("Back from %s", result$id)
      )
      # Change the value of `rv$res` only if
      # the current id is the same as the last_id
      if (result$id == rv$last_id){
        rv$res <- result$res
      }
    }) %...!%
    (function(error){
      warning(error)
    })
  # Note that every render() function should return
  #  something: here it will only work if the
  #  renderPrint() returns a value, even if
  # invisible. We use cat_rule to simulate that.
  cli::cat_rule(
    sprintf("%s sent", rv$last_id)
  )
})


  # output$rnorm will be printed whenever rv$output
  # is available, i.e. returned from the future
  #  and the last one sent.
  output$rnorm <- renderPrint({
    req(rv$res)
  })


  # output$plot will be drawn immediately
  output$plot <- renderPlot({
    plot(iris)
  })
}

shinyApp(ui, server)
```

- We need to treat the outputs in the order they are received. In that case,
  instead of waiting for the very last input, you will need to build a structure

that will receive the output, check if this output is the "next in line", store
it if it is not, or return it if it is, and see if there is another output in the
queue. This type of implementation is a little bit more complex, so we will
not detail a full implementation in this chapter, but here is a small example
of using {liteq} (Csárdi, 2019a).

```r
library(promises)
library(future)
plan(multisession)

library(liteq)
# We create a small db in a tempfile()
temp_queue <- tempfile()
queue <- ensure_queue("jobs", db = temp_queue)
for (i in 1:5){
  future({
    # Faking a random computation time
    Sys.sleep( sample(1:5, 1) )
    return(
      list(
        id = i,
        res = rnorm(5)
      )
    )
  }) %...>%
    # Whenever we receive an output, we add it to
    # the queue database
    (function(results){
      publish(
        queue,
        title = as.character(results$i),
        message = paste(
          results$res,
          collapse = ","
        )
      )
    }) %...!%
    # If ever we have an error, we return it as a warning
    warning(.)
}
Sys.sleep(10)
# List the messages. As you can see, the entries in title
# are not in numerical order because they didn't came back
```

```
# in the same order as they were sent
list_messages(queue)
```

```
  id title status
1  1     3 READY
2  2     4 READY
3  3     2 READY
4  4     1 READY
5  5     5 READY
```

For an example of an application built using {promise} and {future}, feel
free to browse engineering-shiny.org/shinyfuture/[8]: there you will find an example of blocking and non-blocking processes.

---

[8]https://engineering-shiny.org/shinyfuture/

17

Using JavaScript

Prelude

Note you can build a successful, production-grade {shiny}(Chang et al., 2021a) application without ever writing a single line of JavaScript code. Even more when you can use a lot of tools that already bundle JavaScript functionalities: a great example of that being {shinyjs} (Attali, 2020), which allows you to interact with your application using JavaScript, without writing a single line of JavaScript.

We chose to include this chapter in this book as it will help you get a better understanding on how {shiny} works at its core, and show you that getting at ease with JavaScript can help you get better at building web applications using R in the long run. It can also help you extend {shiny} with other JavaScript libraries, for example, using {htmlwidgets} (Vaidyanathan et al., 2020), when you get better at writing JavaScript.

That being said, note also that every inclusion of external JavaScript code or library can present a security risk for your application, so don't include code you don't know/understand in your application unless you are sure of what you are doing. As a rule of thumb, always go for an existing and tested solution when you need JavaScript widgets/functionalities, instead of trying to implement them yourself.

17.1 Introduction

At its core, **building a {shiny} app is building a JavaScript application** that can talk with an R session. This process is invisible to most {shiny} developers, who usually do everything in R. In fact, most of the {shiny} apps out there are 100% written with R.

In fact, when you are writing UI elements in {shiny}, **what you are actually doing is building a series of HTML tags**.

For example, this simple {shiny} (Chang et al., 2021a) code returns a series of HTML tags:

```
fluidPage(
  h2("hey"),
  textInput("act", "Ipt")
)
```

```
<div class="container-fluid">
  <h2>hey</h2>
  <div class="form-group shiny-input-container">
    <label class="control-label" for="act">Ipt</label>
    <input id="act" type="text" class="form-control" value=""/>
  </div>
</div>
```

Later on, when the app is launched, {shiny} binds events to UI elements, and these JavaScript events will communicate with R, in the sense that they will send data to R, and receive data from R. What happens under the hood is a little bit complex and out of scope for this book, but the general idea is that R talks to your browser through a web socket (that you can imagine as a small "phone line" with both software listening at each end, both being able to send messages to the other),[1] and this browser talks to R through the same web socket.

Most of the time, when the JavaScript side of the websocket receives one of these events, the page the user sees is modified (for example, a plot is drawn). On the R end of the websocket, i.e. when R receives data from the web page, a value is fetched, and something is computed.

It's important to note here that the **communication happens in both directions**: from R to JavaScript, and from JavaScript to R. In fact, when we write a piece of code like sliderInput("first_input", "Select a number", 1, 10, 5), what we are doing is creating a binding between JavaScript and R, where the JavaScript runtime (in the browser) listens to any event happening on the slider with the id "plop", and whenever it detects that something happens to this element, something (most of the time its value) is sent back to R, and R does computation based on that value. With output$bla <- renderPlot({}), what we are doing is making the two communicate the other way around: we are telling JavaScript to listen to any

---

[1]See this post on dev.to https://dev.to/buzzingbuzzer/comment/g0g for a quick introduction to the general concept of web sockets.

incoming data from R for the `id "bla"`, and whenever JavaScript sees incoming data from R, it puts it into the proper HTML tag (here, JavaScript inserts the image received from R in the `<img>` tags with the id `bla`).

Even if everything is written in R, we **are** writing a web application, i.e.. HTML, CSS and JavaScript elements. Once you have realized that, the possibilities are endless: in fact almost anything doable in a "classic" web app can be done in `{shiny}` with a little bit of tweaking. What this also implies is that getting (even a little bit) better at writing HTML, CSS, and especially JavaScript will make your app better, smaller, and more user-friendly, as JavaScript is a language that has been designed to interact with a web page: change element appearances, hide and show things, click somewhere, show alerts and prompts, etc. **Knowing just enough JavaScript can improve the quality of your app**: especially when you have been using R to render some complex UIs: think conditional panels, simulating a button click from the server, hide and show elements, etc. All these things are good examples of where you should be using JavaScript instead of building more or less complex `renderUI` or `insertUI` patterns in your server.

Moreover, the number of JavaScript libraries available on the web is tremendous; and the good news is that `{shiny}` has everything it needs to bundle external JavaScript libraries inside your application.[2]

This is what this section of the book aims at: giving you just enough JavaScript knowledge to lighten your `{shiny}` app, in order to improve the global user and developer experience. In this chapter, we will first review some JavaScript basics which can be used "client-side" only, i.e. only in your browser. Then, we will talk about making R and JS communicate with each other, and explore some common patterns for JavaScript in `{shiny}`. Finally, we will quickly present some of the functions available in `{golem}` (Fay et al., 2021) that can be used to launch JavaScript.

Note that this chapter does not try to be a comprehensive JavaScript course. External resources are linked all throughout this chapter and at the end.

---

[2]This can also be done by wrapping a JS libraries inside a package, which will later be used inside an application. See for example `{glouton}` (Fay, 2021e), which is a wrapper around the [js-cookie >https://github.com/js-cookie/js-cookie> JavaScript library.

17.2   A quick introduction to JavaScript

17.2.1   About JavaScript

JavaScript is a programming language which has been designed to work in the browser.[3] To play with a JavaScript console, the fastest way is to open your favorite web browser, and to open the developer tools. In Google Chrome, it's available under View > Developer > Developer Tools. This will open a new interface where you can have access to a JavaScript console under the Console tab. Here, you can try your first JavaScript code! For example, you can try running `var message = "Hello world"; alert(message);`.

As you might have guessed, we will not be focusing on playing with JavaScript in your browser console: what we want to know is how to insert JavaScript code inside a `{shiny}` application.

17.2.2   Including JavaScript code in your app

There are three ways to include the JavaScript code inside your web app:

- As an external file, which is served to the browser alongside your main application page
- Inside a `<script>` HTML tag inside your page
- Inline, on a specific tag, for example by adding an `onclick` event straight on a tag

Note that good practice when it comes to including JavaScript is to add the code inside an external file.

If you are working with `{golem}`, including a JavaScript file is achieved via two functions:

- `golem::add_js_file("name")`, which adds a standard JavaScript file, i.e. one which is not meant to be used to communicate with R. We'll see in the first part of this chapter how to add JavaScript code there.
- `golem::add_js_handler("name")`, which creates a file with a skeleton for `{shiny}` handlers. We'll see this second type of element in the `JavaScript <-> R communication` part.

---

[3]You can now work with JavaScript in a server with Node.JS, but this won't be a useful software when working with `{shiny}`. See linked resources to learn more.

- `golem::add_js_binding("name")`, for more advanced use cases, when you want to build your own custom inputs, i.e. when you want to create a custom HTML element that can be used to interact with {shiny}. See shiny.rstudio.com/articles/js-custom-input.html[4] for more information about how to complete this skeleton.

OK, good, but what do we do now? Note that in this chapter, we will not be covering basic JavaScript object and manipulation. Feel free to refer to the first chapter of JavaScript 4 {shiny} - Field Notes[5] for a detailed introduction to objects and object manipulation, or follow one of the resources linked at the end of this chapter.

17.2.3 Understanding HTML, class, and id

You have to think of a web page as a tree, where the top of the web page is the root node, and every element in the page is a node in this tree (this tree is called a DOM, for Document Object Model). **You can work on any of these HTML nodes with JavaScript**: modify it, bind events to it and/or listen to events, hide and show, etc. But first, **you have to find a way to identify these elements**: either as a group of elements or as a unique element inside the whole tree. That is what HTML semantic elements, classes, and ids are made for. Consider this piece of code:

```
library(shiny)
fluidPage(
  titlePanel("Hello Shiny"),
    textInput("act", "Ipt")
)
```

```
<div class="container-fluid">
  <h2>Hello Shiny</h2>
  <div class="form-group shiny-input-container">
    <label class="control-label" for="act">Ipt</label>
    <input id="act" type="text" class="form-control" value=""/>
  </div>
</div>
```

This {shiny} code creates a piece of HTML code containing three nodes: a

---

[4]https://shiny.rstudio.com/articles/js-custom-input.html
[5]http://connect.thinkr.fr/js4shinyfieldnotes/intro.html

div with a specific class (a Bootstrap container), an h2, which is a level-two
header, and a button which has an id and a class. Both are included in the
div. Let's detail what we have here:

- HTML tags, which are the building blocks of the "tree": here div, h2 and
  button are HTML tags.
- The button has an id, which is short for "identifier". Note that this id has
  to be unique: the id of an element allows you to refer to this exact element.
  In the context of {shiny}, it allows JavaScript and R to talk to each other.
  For example, if you are rendering a plot, you have to be sure it is rendered at
  the correct spot in the UI, hence the need for a unique id in renderPlot().
  Same goes for your inputs: if you are computing a value based on an input
  value, you have to be sure that this value is the correct one.
- Elements can have a class which can apply to multiple elements. This can
  be used in JavaScript, but it is also very useful for styling elements in CSS.

17.2.4   Querying in Vanilla JavaScript

In "Vanilla" JavaScript (i.e. without any external plugins installed), you can
query these elements using methods from the document object. For example:

```
// Given
<div id = "first" name="number" class = "widediv">Hey</div>

// Query with the ID
document.querySelector("#first")
document.getElementById("first")

// With the class
document.querySelectorAll(".widediv")
document.getElementsByClassName("widediv")

// With the name attribute
document.getElementsByName("number")

// Using the tag name
document.getElementsByTagName("div")
```

Note that some of these methods have been introduced with ES6, which is
a version of JavaScript that came out in 2015. This version of JavaScript is
supported by most browsers since mid-2016 (and June 2017 for Firefox) (see

JavaScript Versions[6] from W3Schools). Most of your users should now be using a browser version that is compatible with ES6, but that is something that you might want to keep in mind: browser version matters when it comes to using JavaScript. Indeed, some companies (for internal reason) are still using old versions of Internet Explorer: a constraint you want to be aware of before starting to build the app, hence a question that you want to ask during the Design step.

17.2.5 About DOM events

When users navigate to a web page, they will generate events on the page: clicking, hovering over elements, pressing keys, etc. All these events are listened to by the JavaScript runtime, plus some events that are not generated by the users: for example, there is a "ready" event generated when the web page has finished loading. Most of these events are linked to a specific node in the tree: for example, if you click on something, you are clicking on a node in the DOM. That is where JavaScript events come into play: when an event is triggered in JavaScript, you can link to it a "reaction", in other words a piece of JavaScript code that is executed when this event occurs.

Here are some examples of events:

- `click` / `dblclick`

- `focus`

- `keypress, keydown, keyup`

- `mousedown, mouseenter, mouseleave, mousemove, mouseout, mouseover, mouseup`

- `scroll`

For a full list, please refer to `https://developer.mozilla.org/fr/docs/Web/Events`.

Once you have this list in mind, you can select elements in the DOM, add an **addEventListener** to them, and define a callback function (which is executed when the event is triggered). For example, the code below adds an event to the **input** when a key is pressed, showing a native **alert()** to the user.

---

[6]`https://www.w3schools.com/js/js_versions.asp`

```
<input type="text" id = "firstinput">
<script>
  document.getElementById("firstinput").addEventListener(
    "keypress",
    function(){
      alert("Pressed!")
    }
  )
</script>
```

Note that {shiny} also generates events, meaning that you can customize the behavior of your application based on these events. Here is a code that launches an alert when {shiny} is connected:

```
$(document).on('shiny:connected', function(event) {
  alert('Connected to the server');
});
```

But wait, what is this weird $()? That's jQuery, and we will discover it in the very next section!

17.2.6 About jQuery and jQuery selectors

The jQuery framework is natively included in {shiny}.

---

jQuery is a fast, small, and feature-rich JavaScript library. It makes things like HTML document traversal and manipulation, event handling, animation, and Ajax much simpler with an easy-to-use API that works across a multitude of browsers.

jQuery home page (https://jquery.com)

---

jQuery is a very popular JavaScript library which is designed to manipulate the DOM, its events, and its elements. It can be used to do a lot of things, like hide and show objects, change object classes, click somewhere, etc. And to be able to do that, it comes with the notion of selectors, which will be put between `$()`. You can use, for example:

- `$("#firstinput")` to refer to the element with the id `firstinput`

- `$(".widediv")` to refer to element(s) of class `widediv`

- `$("button:contains('this')")` to refer to the buttons with a text containing `'this'`

You can also use special HTML attributes, which are specific to a tag. For example, the following HTML code:

```
<a href = "https://thinkr.fr" data-value = "panel2">ThinkR</a>
```

contains the `href` and `data-value` attributes. You can refer to these with `[]` after the tag name.

- `$("a[href = 'https://thinkr.fr']")` refers to link(s) with `href` being `https://thinkr.fr`

- `$('a[data-value="panel2"]')` refers to link(s) with `data-value` being `"panel2"`

These and other selectors are **used to identify one or more node(s) in the big tree which is a web page**. Once we have identified these elements, we can either extract or change data contained in these nodes, or invoke methods contained within these nodes. Indeed JavaScript, like R, can be used as a functional language, but most of what we do is done in an object-oriented way. In other words, you will interact with objects from the web page, and these objects will contain data and methods.

Note that this is not specific to `jQuery`: elements can also be selected with standard JavaScript. `jQuery` has the advantage of simplifying selections and actions and is a cross-platform library, making it easier to ship applications that can work on all major browsers. And it comes with `{shiny}` for free!

Choosing `jQuery` or vanilla JavaScript is up to you: and in the rest of this chapter we will try to mix both syntaxes, and put both when possible, so that you can choose the one you are the most comfortable with.

17.3 Client-side JavaScript

It is hard to give an exhaustive list of what you can do with JavaScript inside {shiny}. As a {shiny} app is part JavaScript, part R, once you have a good grasp of JavaScript you can quickly enhance any of your applications. That being said, a few common things can be done that would allow you to immediately optimize your application: i.e. small JavaScript functions that will prevent you from writing complex algorithmic logic in your application server.

17.3.1 Common patterns

- `alert("message")` uses the built-in alert-box mechanism from the user's browser (i.e., the `alert()` function is not part of `jQuery` but it is built inside the user's browser). It works well as it relies on the browser instead of relying on R or on a specific JavaScript library. You can use this functionality to replace a call to {shinyalert} (Attali and Edwards, 2020): the result is a little less aesthetically pleasing, but is easier to implement and maintain.

- `var x = prompt("this", "that");` this function opens the built-in prompt, which is a text area where the user can input text. With this code, when the user clicks "OK", the text is stored in the x variable, which you can then send back to R (see later in this chapter for more info on how to do that). This can replace something like the following:

```r
# Initiating a modalDialog that will ask the user to enter
# some information
mod <- function() {
  # The modal box definition
  modalDialog(
    # Simple body with a textInput
    tagList(
      textInput("info", "Your info here")
    ),
    footer = tagList(
      modalButton("Cancel"),
      actionButton("ok", "OK")
    )
  )
}
```

```
# When the user clicks on the "show" button in the UI,
# the modalDialog() is displayed
observeEvent(input$show, {
  showModal(mod())
})

# Whenever the "ok" button is clicked, the modal is removed
observeEvent(input$ok, {
  print(input$info)
  removeModal()
})
```

- `$('#id').css('color', 'green');,` or in vanilla JavaScript `document.getElementById("demo").style.color = "green";` changes the CSS attributes of the selected element(s). Here, we are switching to green on the #id element.

- `$("#id").text("this"),` or in vanilla JavaScript `document.getElementById("id").innerText = "this";` changes the text content to "this". This can be used to replace the following:

```
output$ui <- renderUI({
  # Conditionnal rendering of the UI
  if (this){
    tags$p("First")
  } else {
    tags$p("Second")
  }
})
```

- `$("#id").remove();,` or in vanilla JavaScript `var elem = document.querySelector('#some-element'); elem.parentNode.removeChild(elem);` completely removes the element from the DOM. It can be used as a replacement for `shiny::removeUI()`, or as a conditional UI. Note that this code doesn't remove the input values on the server side: the elements only disappear from the UI, but nothing is sent to the server side. For a safe implementation, see {shinyjs}.

17.3.2 Where to put them: Back to JavaScript Events

OK, now that we have some ideas about JS code that can be used in {shiny}, where do we put it? HTML and JS have a concept called **events**, which are, well, events that happen when the user manipulates the web page: when the user clicks, hovers (the mouse goes over an element), presses keys on the keyboard, etc. All these events can be used to trigger a JavaScript function.

Here are some examples of adding JavaScript functions to DOM events:

- `onclick`

The `onclick` attribute can be added straight inside the HTML tag when possible:

```r
# Building a button using the native HTML tag
# (i.e. not using the actionButton() function)
# This button only goal is to launch this JS code
# when it is clicked
tags$button(
  "Show",
  onclick = "$('#plot').show()"
)
```

Or with `shiny::tagAppendAttributes()`:

```r
# Using tagAppendAttributes() allows to add attributes to the
# outputed UI element
plotOutput(
  "plot"
) %>% tagAppendAttributes(
  onclick = "alert('hello world')"
)
```

Here is, for example, a small {shiny} app that implements this behavior:

```r
library(shiny)
library(magrittr)
ui <- function(){
  fluidPage(
    # We create a plotOutput, which will show an alert when
    # it is clicked
    plotOutput(
      "plot"
```

```
    ) %>% tagAppendAttributes(
        onclick = "alert('iris plot!')"
    )
  )
}

server <- function(input, output, session){
  output$plot <- renderPlot({
    plot(iris)
  })
}

shinyApp(ui, server)
```

You can find a real-life example of this `tagAppendAttributes` in the {tidytuesday201942} (Fay, 2021k) app:

- R/mod_dataviz.R#L109[7], where clicking the plot generates the creation of a {shiny} input (we will see this below)

That, of course, works well with very short JavaScript code. For longer JavaScript code, you can write a function inside an external file, and add it to your app. In {golem}, this works by launching the `add_js_file("name")`, which will create a `.js` file. The JavaScript file is then automatically linked in your application.

This, for example, could be:

- In `inst/app/www/script.js`

```
function alertme(id){
  // Asking information
  var name = prompt("Who are you?");
  // Showing an alert
  alert("Hello " + name + "! You're seeing " + id);
}
```

- Then in R

---

[7]https://github.com/ColinFay/tidytuesday201942/blob/master/R/mod_dataviz.R# L109

```
plotOutput(
  "plot"
) %>% tagAppendAttributes(
  # Calling the function which has been defined in the
  # external script
  onclick = "alertme('plot')"
)
```

Inside this `inst/app/www/script.js`, you can also attach a new behavior with jQuery to one or several elements. For example, you can add this `alertme` / `onclick` behavior to all plots of the app:

```
function alertme(id){
  var name = prompt("Who are you?");
  alert("Hello " + name + "! You're seeing " + id);
}

/* We're adding this so that the function is launched only
when the document is ready */
$(function(){
  // Selecting all `{shiny}` plots
  $(".shiny-plot-output").on("click", function(){
    /* Calling the alertme function with the id
    of the clicked plot */
    alertme(this.id);
  });
});
```

Then, all the plots from your app will receive this on-click event.[8]

Note that there is a series of events which are specific to {shiny}, but that can be used just like the one we have just seen:

```
// We define a function that will ask for the visitor name, and
// then show an alert to welcome the visitor
function alertme(){
  var name = prompt("Who are you?");
  alert("Hello " + name + "! Welcome to my app");
```

---

[8]This `click` behavior can also be done through `$(".shiny-plot-output").click(...)`. We chose to display the `on("click")` pattern as it can be generalized to all DOM events.

```
}

$(function(){
  // Waiting for `{shiny}` to be connected
  $(document).on('shiny:connected', function(event) {
    alertme();
  });
});
```

See JavaScript Events in {shiny}[9] for the full list of JavaScript events available in {shiny}.

17.4 JavaScript <-> {shiny} communication

Now that we have seen some client-side optimization, i.e. R does not do anything with these events when they happen (in fact R is not even aware they happened), let's now see how we can make these two communicate with each other.

17.4.1 From R to JavaScript

Calling JS from the server side (i.e. from R) is done by defining a series of `CustomMessageHandler` functions: these are functions with one argument that can then be called using the `session$sendCustomMessage()` method from the server side. Or if you are using {golem}, using the `invoke_js()` function.

You can define them using this skeleton:

```
$( document ).ready(function() {
  Shiny.addCustomMessageHandler('fun', function(arg) {

  })
});
```

This skeleton is the one from `golem::add_js_handler("first_handler")`.

Then, it can be called from server side with:

---

[9]https://shiny.rstudio.com/articles/js-events.html

```r
session$sendCustomMessage("fun", list())
# OR
golem::invoke_js("fun", ...)
```

Note that the `list()` argument from your function will be converted to JSON, and read as such from JavaScript. In other words, if you have an argument called x, and you call the function with `list(a = 1, b = 12)`, then in JavaScript you will be able to use `x.a` and `x.b`.

For example:

- In `inst/app/www/script.js`:

```js
// We define a handler called "computed", that can be called
// from the server side of the {shiny} application
Shiny.addCustomMessageHandler('computed', function(mess) {
  // The received value (in mess) is serialized in JSON,
  // so we can  access the list element with object.name
  alert("Computed " + mess.what + " in " + mess.sec + " secs");
})
```

- Then in R:

```r
observe({
  # Register the starting time
  deb <- Sys.time()
  # Mimic a long computation
  Sys.sleep(
    sample(1:5, 1)
  )
  # Calling the computed handler
  golem::invoke_js(
    "computed",
    # We send a list, that will be turned into JSON
    list(
      what = "time",
      sec = round(Sys.time() - deb)
    )
  )
})
```

17.4.2   From JavaScript to R

How can you do the opposite (from JavaScript to R)? {shiny} apps, in the browser, contain an object called `Shiny`, which may be used to send values to R by creating an `InputValue`. For example, with:

```
// This function from the Shiny JavaScript object
// Allows to register an input name, and a value
Shiny.setInputValue("rand", Math.random())
```

you will bind an input that can be caught from the server side with:

```
# Once the input is set, it can be caught with R using:
observeEvent( input$rand , {
  print( input$rand )
})
```

`Shiny.setInputValue`[10] can, of course, be used inside any JavaScript function. Here is a small example that wraps up some of the things we have seen previously:

- In `inst/app/www/script.js`

```
function alertme(){
  var name = prompt("Who are you?");
  alert("Hello " + name + "! Welcome to my app");
  Shiny.setInputValue("username", name)
}

$(function(){
  // Waiting for `{shiny}` to be connected
  $(document).on('shiny:connected', function(event) {
    alertme();
  });

  $(".shiny-plot-output").on("click", function(){
    /* Calling the alertme function with the id
```

---

[10]Note that the old name of this function is `Shiny.onInputChange`.

```
    of the clicked plot.
    The `this` object here refers to the clicked element*/
    Shiny.setInputValue("last_plot_clicked", this.id);
  });
});
```

These events (getting the user name and the last plot clicked), can then be caught from the server side with:

```
# We wait for the output of alertme(), which will set the
# "username" input value
observeEvent( input$username , {
  cli::cat_rule("User name:")
  print(input$username)
})
```

```
# This will print the id of the last clicked plot
observeEvent( input$last_plot_clicked , {
  cli::cat_rule("Last plot clicked:")
  print(input$last_plot_clicked)
})
```

Which will give:

```
> golex::run_app()
Loading required package: shiny

Listening on http://127.0.0.1:5495
  User name:
[1] "Colin"
  Last plot clicked:
[1] "plota"
  Last plot clicked:
[1] "plotb"
```

Important note: If you are using modules, you will need to pass the namespacing of the `id` to be able to get it back from the server. This can be done using the `session$ns` function, which comes by default in any golem-generated module. In other words, you will need to write something like the following:

```
$( document ).ready(function() {
  // Setting a custom handler that will
  // ask the users their name
  // then set the returned value to a Shiny input
  Shiny.addCustomMessageHandler('whoareyou', function(arg) {
    var name = prompt("Who are you?")
    Shiny.setInputValue(arg.id, name);
  })
});


mod_my_first_module_ui <- function(id){
  ns <- NS(id)
  tagList(
    actionButton(
      ns("showname"), "Enter your name"
    )
  )
}

mod_my_first_module_server <- function(input, output, session){
  ns <- session$ns
  # Whenever the button is clicked,
  # we call the CustomMessageHandler
  observeEvent( input$showname , {
    # Calling the "whoareyou" handler
    golem::invoke_js(
      "whoareyou",
      # The id is namespaced,
      # so that we get it back on the server-side
      list(
        id = ns("name")
      )
    )
  })

  # Waiting for input$name to  be set with JavaScript
  observeEvent( input$name , {
    cli::cat_rule("Username is:")
    print(input$name)
  })
}
```

17.5   About {shinyjs} JS functions

As said in the introduction to this chapter, running JavaScript code that you
don't fully control/understand can be tricky and might open doors for external
attacks. In many cases, for the most common JavaScript manipulations, it's
safer to go for a package that has already been proved efficient: {shinyjs}.

This package, licensed in MIT since version 2.0.0, can be used to perform
common JavaScript tasks: show, hide, alert, click, etc.

See deanattali.com/shinyjs/[11] for more information about how to use this
package.

17.6   One last thing: API calls

If your application uses API calls, chances are that right now you have been
doing them straight from R. But there are downsides to that approach. Notably, if the API limits requests based on an IP and your application gets a
lot of traffic, your users will end up being unable to use the app because of
this restriction.

So, why not switch to writing these API calls in JavaScript? As JavaScript is
run inside the user's browser, the limitation will apply to the user's IPs, not
the one where the application is deployed, allowing you to more easily scale
your application.

You can write this API call using the `fetch()` JavaScript function. It can
then be used inside a {shiny} JavaScript handler, or as a response to a
DOM event (for example, with `tags$button("Get Me One!", onclick =
"get_rand_beer()")`, as we will see below).

Here is the general skeleton that would work when the API does not need
authentication and returns JSON.

- Inside JavaScript (here, we create a function that will be available on an
  `onclick` event)

---

[11]https://deanattali.com/shinyjs/

```
// FUNCTION definition
const get_rand_beer = () => {
  // Fetching the data
  fetch("https://api.punkapi.com/v2/beers/random")
  // What do we do when we receive the data
  .then((data) =>{
  // TTurn the data to JSON
    data.json().then((res) => {
    // Send the json to R
      Shiny.setInputValue("beer", res, {priority: 'event'})
    })
  })
  // Define what happens if we fail to fetch
  .catch((error) => {
      alert("Error catching result from API")
  })
};
```

- Observe the event in your server:

```
observeEvent( input$beer , {
  # Do things with beer
})
```

Note that the data shared between R and JavaScript is serialized to JSON, so you will have to manipulate that format once you receive it in R.

Learn more about `fetch()` at Using Fetch[12].

17.7 Learn more about JavaScript

If you want to interact straight from R with NodeJS (JavaScript in the terminal), you can try the {bubble} (Fay, 2021a) package. Be aware that you will need to have a working NodeJS installation on your machine.

It can be installed from GitHub

---

[12]https://developer.mozilla.org/en-US/docs/Web/API/Fetch_API/Using_Fetch

```r
remotes::install_github("ColinFay/bubble")
```

You can use it in RMarkdown chunks by setting the {knitr} engine:

```r
bubble::set_node_engine()
```

Or straight in the command line with:

```r
node_repl()
```

Want to learn more? Here is a list of external resources to learn more about JavaScript:

17.7.1 {shiny} and JavaScript

- We have written an online, freely available book about {shiny} and JavaScript: *JavaScript 4 {shiny} - Field Notes*[13].

- JavaScript for {shiny} Users[14], companion website to the rstudio::conf(2020) workshop.

- Build custom input objects[15].

- Packaging JavaScript code for {shiny}[16].

- Communicating with {shiny} via JavaScript[17].

17.7.2 JavaScript basics

- Mozilla JavaScript[18]
- w3schools JavaScript[19]
- Free Code Camp[20]

---

[13] http://connect.thinkr.fr/js4shinyfieldnotes/
[14] https://js4shiny.com/
[15] https://shiny.rstudio.com/articles/building-inputs.html
[16] https://shiny.rstudio.com/articles/packaging-javascript.html
[17] https://shiny.rstudio.com/articles/communicating-with-js.html
[18] https://developer.mozilla.org/en-US/docs/Web/JavaScript
[19] https://www.w3schools.com/js/default.asp
[20] https://www.freecodecamp.org/

- JavaScript for Cats[21]
- Learn JS[22]

17.7.3 jQuery

- jQuery Learning Center[23]
- w3schools jQuery[24]

17.7.4 Intermediate/advanced JavaScript

- Eloquent JavaScript[25]
- You Don't Know JS Yet[26]

---

[21]http://jsforcats.com/
[22]https://www.learn-js.org/
[23]https://learn.jquery.com/
[24]https://www.w3schools.com/jquery/default.asp
[25]https://eloquentjavascript.net/
[26]https://github.com/getify/You-Dont-Know-JS

18

A Gentle Introduction to CSS

18.1 What is CSS?

18.1.1 About CSS

CSS, for `Cascading Style Sheets`, is one of the main technologies that powers the web today, along with HTML and JavaScript. HTML is a series of tags that define your web page structure, and JavaScript is a programming language that allows you to manipulate the page (well, it can do a lot more than that, but we are simplifying to make it understandable). **CSS is what handles the design, i.e. the visual rendering of the web page: the color of the header, the font, the background, and everything that makes a web page look like it is not from 1983** (again, we are simplifying for the sake of clarity). In every browser, each HTML element has a default style: for example, all `<h1>` have the size `2em` and are in bold, and `<strong>` is in bold. But we might not be happy with what a "standard page" (with no CSS) looks like: that is the very reason for CSS, modifying the visual rendering of the page.

If you want to get an idea of the importance of CSS, try installing extensions like Web Developer[1] for Google Chrome. Then, if you go on the extension and choose CSS, click "Disable All Style", to see what a page without CSS looks like.

For example, Figure 18.1 is what rtask.thinkr.fr[2] looks like with CSS, and Figure 18.2 and Figure 18.3 shows what it looks like without CSS.

CSS now seems pretty useful right?

---

[1]https://chrome.google.com/webstore/detail/web-developer/
bfbameneiokkgbdmiekhjnmfkcnldhhm
[2]https://rtask.thinkr.fr

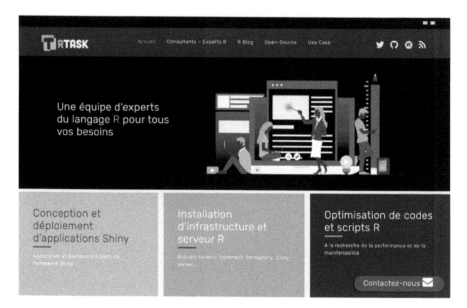

FIGURE 18.1: `https://rtask.thinkr.fr` with CSS.

FIGURE 18.2: `https://rtask.thinkr.fr` without CSS.

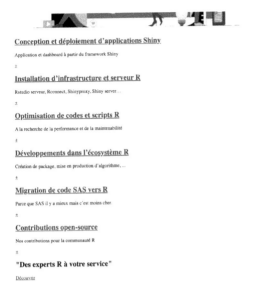

FIGURE 18.3: `https://rtask.thinkr.fr` without CSS.

18.1.2   {shiny}'s default: `fluidPage()`

In {shiny}, there is a default CSS: the one from Bootstrap 3. As you can see, if you have created a `fluidPage()` before, there is already styling applied.

Compare:

- No `fluidPage`, Figure 18.4

```r
library(shiny)
ui <- function(){
  tagList(
    h1("Hey"),
    h2("You"),
    p("You rock!"),
    selectInput("what", "Do you", unique(iris$Species))
  )
}

server <- function(
  input,
  output,
```

```
    session
){

}

shinyApp(ui, server)
```

FIGURE 18.4: {shiny} without CSS.

- With fluidPage, Figure 18.5

```
library(shiny)
ui <- function(){
  fluidPage(
    h1("Hey"),
    h2("You"),
    p("You rock!"),
    selectInput("what", "Do you", unique(iris$Species))
  )
}
```

```
server <- function(
  input,
  output,
  session
){

}

shinyApp(ui, server)
```

FIGURE 18.5: {shiny} with its default CSS.

Yes, that is subtle, but you can see how it makes the difference on larger apps.

18.2 Getting started with CSS

CSS is a descriptive language, meaning that you will have to declare the style either on a tag or inside an external file. We will see how to integrate CSS inside your {shiny} application in the next section, but before that, let's start with a short introduction to CSS.[3]

18.2.1   About CSS syntax

CSS syntax is composed of two elements: a selector and a declaration block. The CSS selector describes how to identify the HTML tags that will be affected by the style declared with key-value pairs in the declaration block that follows. And because an example will be easier to understand, here is a simple CSS rule:

```
h2 {
  color:red;
}
```

Here, the selector is h2, meaning that the HTML tags aimed by the style are the <h2> tags. The declaration block contains the key-value pair telling that the color will be red. Note that each key-value pair must end with a semicolon.

18.2.2   CSS selectors

CSS selectors are a wide topic, as there are many combinations of things you might want to select inside an HTML page.

The first selector types are the "standard" ones: name, id, or class. These refer to the elements composing an HTML tag: for example, with <h2 id = "tileone" class = "standard">One</h2>, the name is h2, the id is tileone, and the class is standard.[4]

To select these three elements in CSS:

- Write the name as-is: h2.
- Prefix the id with #: #tileone.
- Prefix the class with .: .standard.

You can also combine these elements: for example, h2.standard will select all the h2 tags with a class standard, and h2,h3 will select h2 and h3.

---

[3]Of course, this part will not make you an expert CSS programmer, but we hope you will get an idea of how it works, enough to get you started and want to learn more!

[4]Note that in HTML, id must be unique, but class must not.

You can build more complex selectors: for example `div.standard > p` will select all the `<p>` tags that are contained inside a `div` of class `standard` (CSS combinator), or `a:hover`, which dictates the style of the `a` tags when they are hovered by the mouse (CSS pseudo-class), `div.standard::first-letter`, which selects the first letter of the `div` of class `standard` (CSS pseudo-elements), and `h2[data-value="hey"]`, which selects all the `h2` with a `data-value` attribute set to `"hey"` (CSS attribute selector).

As you can see, lots of complex selectors can be built with CSS, to target very specific elements of your UI. But mastering these complex selectors is not the main goal of this chapter, hence we will just be using standard selectors in the rest of the examples in this book.

18.2.3 CSS properties

Now that you have selected elements, it is time to apply some styles! Between the brackets of the declaration block, you will have to define a series of key-value elements defining the properties of the style: the key here is the CSS property, followed by its value.

For example, `color: red;` or `text-align: center;` defines that for the selected HTML elements, the color will be red, or the text centered. We will not cover all the possible properties, as there are hundreds of them. Feel free to refer to the CSS Reference[5] page from Mozilla for an exhaustive list of available properties.

18.3 Integrate CSS files in your {shiny} app

Now that you have an idea of how to start writing your own CSS, how do you integrate it into your {shiny} application? There are three methods that can be used: writing it inline, integrating it inside a `tags$script()` straight into your application UI code, or by writing it into an external file. Note that good practice is considered to be the integration of an external file.

18.3.1 Inline CSS

If you need to add style to one specific element, you can write it straight inside the HTML tag, as seen in Figure 18.6:

---

[5]`https://developer.mozilla.org/en-US/docs/Web/CSS/Reference`

```
library(shiny)
ui <- function(){
  tagList(
    h2(style = "color:red;", "This is red")
  )
}

server <- function(
  input,
  output,
  session
){

}

shinyApp(ui, server)
```

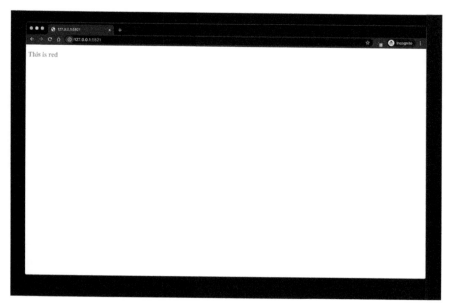

FIGURE 18.6: {shiny} with a red title.

But this method loses all the advantages of CSS, notably, the possibility of applying style to multiple elements. Use it with caution.

18.3.2 Writing in a `tags$style()`

If you had a `tags$style()` somewhere inside your UI code (generally at the very beginning of your UI), you could then add CSS code straight to your application.

Figure 18.7 is an example:

```
library(shiny)
ui <- function(){
  tagList(
    tags$style(
      "h2{
        color:red;
      }"
    ),
    h2("This is red")
  )
}

server <- function(
  input,
  output,
  session
){

}

shinyApp(ui, server)
```

This works, but should not be considered as the best option: indeed, if you have a large amount of CSS code to insert in your app, it can make the code harder to read as it adds a large amount of visual noise.

The best solution, then, is to go with the alternative of writing the CSS inside a separate file: it allows you to separate things and to make the UI code smaller, as it is easier to maintain a separate CSS file than a CSS written straight into R code.

18.3.3 Including external files

To include an external CSS file, you will have to use another tag: `tags$link()`. What this tag will contain is these three elements:

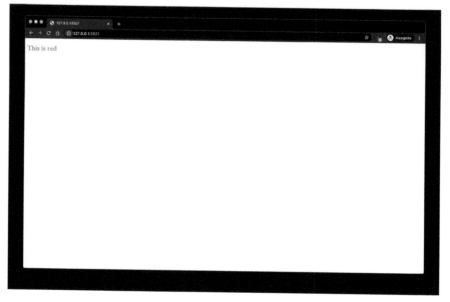

FIGURE 18.7: {shiny} with a red title.

- rel="stylesheet"
- type="text/css"
- href="www/custom.css"

The first two are standard: you do not need to change them; they are necessary to indicate to the HTML page that you are creating a stylesheet, with the type being text/css. The href is the one you will need to change: this path points to where your style file is located.

If you are building your application with {golem} (Fay et al., 2021), the good news is that this file creation and linking is transparent: if you call golem::add_css_file("name"), a file will be created at inst/app/www, and this file will be automatically linked inside your UI thanks to the bundle_resources() function.

18.3.4   Using R packages

If you want to use an external CSS template, there are several packages that exist that can implement new custom UI designs for your application. Here are some:

- **{resume}**(Fay, 2021i), provides an implementation of the Bootstrap Resume Template[6].

- **{nessy}** (Fay, 2021h), a port of NES CSS[7].

- **{skeleton}** (Fay, 2021j), Skeleton CSS[8].

- **{shinyMobile}** (Granjon et al., 2021), shiny API for Framework7 (IOS/android).

- **{shinydashboardPlus}** (Granjon, 2021b), extensions for shinydashboard.

- **{bs4Dash}** (Granjon, 2021a), Bootstrap 4 shinydashboard using AdminLTE3: an example is available at engineering-shiny.org/bs4dashdemo/[9]

- **{fullPage}**(Coene, 2020), fullPage.js, pagePiling.js, and multiScroll.js for shiny.

And all the amazing things done at RinteRface[10].

18.4   External resources

If you want to learn more about CSS, there are three places where you can get started:

- FreeCodeCamp[11], which contains many course hours covering HTML and CSS.

- W3 Schools CSS Tutorial[12]

- Learn to style HTML using CSS[13]

---

[6]https://github.com/BlackrockDigital/startbootstrap-resume
[7]https://github.com/nostalgic-css/NES.css
[8]http://getskeleton.com/
[9]https://engineering-shiny.org/bs4dashdemo/
[10]https://github.com/RinteRface
[11]https://www.freecodecamp.org/learn
[12]https://www.w3schools.com/css/
[13]https://developer.mozilla.org/en-US/docs/Learn/CSS

Part VIII

Appendix

Appendix A - Use Case: Building an App from Start to Finish

This chapter aims at exemplifying the workflow developed in this book using a "real-life" example. In this appendix, we will be building a {shiny} application from start to finish. We've chosen to build an application that doesn't rely on heavy computation/data analysis, so that we can focus on the engineering process, not on the internals of the analytic methodology, nor on spending time explaining the dataset.

About the application

In this appendix, we will build a "minify" application, an application that takes a CSS, JavaScript, HTML or JSON input, and outputs a minified version of this file. We will rely on the {minifyr} package to build this application.

Here is an example of what the specifications for this app could look like:

Hello!

We want to build a small application that can
minify CSS, JavaScript, HTML and JSON.

In this app, user will be able either to paste
the content or to upload a file.

Once the content is pasted/upladed, they select
the type, which is pre-selected based on the
file extension. Then they click on a button,
and the content is minified.

They can then copy the output, or download it as a file.

```
Cheers!
```

Step 1: Design

Deciphering the specifications

General observations

- As this app is pretty straightforward, it would be better to handle everything in the same page, *i.e* everything should happen on the same page (no tab).

- It would be a plus to have the "before minification"/"after minification" gain, so that the users have a better idea of the purpose of the application.

User experience considerations

- We should provide a link to an explanation of minification.

- The user might get different results based on the minifying algorithm they use, which can be surprising at first. The application should alert about this.

- For long printed outputs, if we use `verbatimTextOuput`, we should be careful about the page width, as these elements will natively overflow on the x-axis of the page. This should be doable with the following CSS: `pre{ white-space: pre-wrap; word-break: keep-all; }`.

- We should be careful about using semantic HTML for the inputs and outputs.

Technical points

- As {minifyr} wraps a NodeJS module, we will need to install NodeJS when deploying.

- To be sure that the process works, we should check the validity of the file extension from the UI and from the server side.

Building a concept map

Figure 18.8 is the concept map for this application, using Xmind.

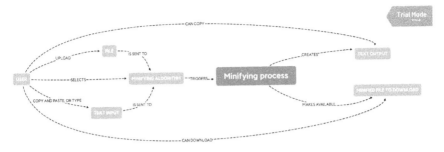

FIGURE 18.8: A concept map for the minifying application.

Asking questions

About the end users

- *Who are the end users of your app?*

This application is mainly useful for web developers.

- *Are they tech-literate?*

Yes.

- *In which context will they be using your app?*

Notably at work, or while building pet projects.

- *On what machines?*

Laptop/personal computer. Small chance of using this on a smartphone.

- *What browser version will they be using?*

Hard to say, but given that we aim for a public of community developers, probably modern browsers.

- *Will they be using the app in their office, on their phone while driving a tractor, in a plant, or while wearing lab coats?*

Nothing of the like: they should be using this application while developing, so chances are they are using it at a desk.

Building personas

Let's pick two random names for our personas, and two fake companies where they might be working.

```
nms <- withr::with_seed(
  608, {
    charlatan::ch_name(2)

  }
)
nms
```

```
[1] "Delina Stanton" "Theo Torphy"
```

```
company <- withr::with_seed(
  608, {
    charlatan::ch_company(2)
  }
)
company
```

```
[1] "Witting-Witting" "Stehr-Stehr"
```

Delina Stanton: {shiny} developer at Witting-Witting

Delina Stanton is a {shiny} developer at Witting-Witting. She's been learning R in graduate school while working on her master's degree in statistics. When she started at Witting-Witting, she was mainly doing data analysis in Rmd, but has gradually switched to building {shiny} applications full-time. She discovered minification while reading the "Engineering Production-Grade Shiny App" book, and now wants to add this to her {shiny} application.

Theo Torphy: web developer and trainer at Stehr-Stehr

Theo Torphy is a web developer at Stehr-Stehr. He studied web development at the university, where he learned about minification. He is now also in charge of training new recruits for the company where he works, and also gives some lectures at the university he attended. Most of the minification he does is automated, but he is looking for a tool he can use during training and classes to explain how minification works.

This step is available at `https://github.com/ColinFay/minifying/tree/master/step-1-design`.

Step 2: Prototyping

In this step, we will be building the back-end of the application on one side, and the UI on the other side. Once we have the back-end settled and the UI defined, we will be working on making the two work with each other.

Back-end in Rmd

Our back-end will be composed of two functions:

- `guess_minifier`, which will take a function and return the available algorithms for that file: for example, if you have a JavaScript file, you'll be able to use the `minifyr_js_babel()`, `minifyr_js_gcc()`, `minifyr_js_terser()`, `minifyr_js_uglify()`, and `minifyr_js_yui()` functions. If the type is not guessed based on the extension, the function should fail gracefully, and not make {shiny} crash. We'll chose to return an empty string if this extension is not guessed.

```r
library(minifyr)
guess_minifier <- function(file){
  # We'll start by getting the file extension
  ext <- tools::file_ext(file)
  # Check that the extension is correct, if not, return early
  # It's important to do this kind of check also
  # on the server side as HTML manual tempering
  # would allow to also send other type of files
  if (
    ! ext %in% c("js", "css", "html", "json")
  ){
    # Return early
    return(list())
  }
```

```r
# We'll then retrieve the available
# pattern based on the extension
patt <- switch(
  ext,
  js = "minifyr_js_.+",
  html = "minifyr_html_.+",
  css = "minifyr_css_.+",
  json = "minifyr_json_.+"
)
# List all the available functions to minify the file
list(
  file = file,
  ext = ext,
  # We return this pattern so that
  # it will be used to update the selectInput that
  # is used to select an algo
  pattern = patt,
  functions = grep(
    patt,
    names(
      loadNamespace("minifyr")
    ),
    value = TRUE
  )
)

}


# minifyr comes with a series of examples,
# so we can use them as tests
guess_minifier(
  minifyr_example("css")
)[2:4]


$ext
[1] "css"

$pattern
[1] "minifyr_css_.+"

$functions
[1] "minifyr_css_cleancss" "minifyr_css_cssnano"
[3] "minifyr_css_csso"     "minifyr_css_sqwish"
[5] "minifyr_css_yui"      "minifyr_css_crass"
```

```
guess_minifier(
  minifyr_example("js")
)[2:4]


$ext
[1] "js"

$pattern
[1] "minifyr_js_.+"

$functions
[1] "minifyr_js_terser" "minifyr_js_yui"
[3] "minifyr_js_gcc"    "minifyr_js_babel"
[5] "minifyr_js_uglify"


guess_minifier(
  minifyr_example("html")
)[2:4]


$ext
[1] "html"

$pattern
[1] "minifyr_html_.+"

$functions
[1] "minifyr_html_minifier"


guess_minifier(
  minifyr_example("json")
)[2:4]


$ext
[1] "json"

$pattern
[1] "minifyr_json_.+"

$functions
[1] "minifyr_json_jsonminify"
```

```
# Try with a non valid extension
guess_minifier(
  "path/to/text.docx"
)
```

```
list()
```

- A compress() function, which takes three parameters: the file as input, the algo outputted from our last function, and the selection, which is the one selected by the user. The compressed file will be outputted to a tempfile.

```
compress <- function(algo, selection){
  # Creating a tempfile using our algo object
  tps <- tempfile(fileext = sprintf(".%s", algo$ext))
  # Getting the function with the selection
  converter <- get(
    grep(selection, algo$functions, value = TRUE)
  )
  # Do the conversion
  converter(algo$file, tps)
  # Return the temp file
  return(tps)
}
```

```
algo <- guess_minifier(
  minifyr_example("js")
)
```

```
compress(
  algo = algo,
  selection = "babel"
)
```

- Finally, a compare() function, that can compare the size of two files, so that we can measure the minification gain. This function will take two file paths.

```
compare <- function(original, minified){
  # Get the file size of both
```

```
  original <- fs::file_info(original)$size
  minified <- fs::file_info(minified)$size
  return(original - minified)
}
```

So, for the complete process:

```
algo <- guess_minifier(
  minifyr_example("js")
)

compressed <- compress(
  algo = algo,
  selection = "babel"
)
compare(
  minifyr_example("js"),
  compressed
)
```

375

Now, time to move this into a vignette!

UI prototyping

Let's start by drawing a small mockup of our front-end using Excalidraw[14], as seen in Figure 18.9.

We would love this application to be "full screen", and to do that, we'll take inspiration from the split-screen layout[15] available at W3Schools. To mock up the UI, we will also use the {shinipsum} package.

In this first step, we will start generating the module skeleton for the application. Here, we will have a left module for the left part of the app, and a right module for the right part. Each will receive their corresponding class, based on the CSS from W3. Now that these two spots are available, we'll add the two modules, with some fake output, to simulate our application behavior. The left side will be functional in the sense that uploading a file will

---

[14]https://excalidraw.com/
[15]https://www.w3schools.com/howto/howto_css_split_screen.asp

FIGURE 18.9: A mockup for the UI of our application, made with Excalidraw https://excalidraw.com/.

randomly add algorithms to the `selectInput()`, and clicking on `Launch the minification` will regenerate a fake text.

Now, let's pick a soft palette of colors, using coolors.co[16], and a font family from fonts.google.com[17]. We went for:

- One of the monochrome palettes from coolors.co[18].
- The `Sora` font fonts.google.com/specimen/Sora[19]. There is not that much text displayed on the screen, so this font should work well.

We then used CSS to arrange our page: size, padding, alignment, colors, etc. If you want to know more about this file, it's located in the `inst/app/www` folder of the package.

This step is available at github.com/ColinFay/minifying, folder step-2-prototype[20].

---

[16]https://coolors.co/palettes/
[17]https://fonts.google.com/
[18]https://coolors.co/fbfbf2-e5e6e4-cfd2cd-a6a2a2-847577
[19]https://fonts.google.com/specimen/Sora
[20]https://github.com/ColinFay/minifying/tree/master/step-2-prototype

Step 3: Build

Now we've got the back-end in an Rmd, the front end working with {shinipsum}. Now is the time to make the two work together!

Here is the logic we will be adding to the application:

- When a file is uploaded, we also check the format from the server-side: the UI restriction with the `accept` parameter of `fileInput()` will not be enough to stop users who **really** want to upload something else.

- If the file comes with the right extension, we update the algorithm selection and read it inside the "Original content" block.

- Once the user clicks on "Launch the minification", we create a temp file and minify the original file inside this temp file.

- When the file is minified, we update the gain output to reflect how many bytes have been gained from the minification, and add the result of this minification to the "Minified content" block.

- Finally, when the "Download the output" button is clicked, the minified file is downloaded.

During this process, we will migrate the functions from the Rmd to files inside the `R/` folder, use external dependencies, and document our business logic functions. You can refer to the `dev/02_dev.R` file if you want to read the exact steps taken here.

This step is available at `https://github.com/ColinFay/minifying/tree/master/step-3-build`.

Step 4: Strengthen

As of now, we have a working application. Time to strengthen it!

Here are the few steps we will be working on:

- Turning our business logic into an R6 class. This R6 class will generate an object at the very start of our app, and it will be passed into the modules.

- As the minification process takes a couple of seconds, we will add a small progress bar so that the user knows something is happening.

- As we will use R6, we will need to manually set the reactive context invalidation. To do so, we will use `triggers` from {gargoyle}.

- Chances are, users will be testing several algorithms when using the application, and we don't want the minification process to happen another time when it is called on the same file and with the same algorithm. This is even even more important because the process involves calling an external NodeJS process. To prevent that, we will be caching the function that does the computation.

- Create an unseen input that will upload data, so that we can build an interactivity test using {crrry}. This input will look like the following on the server:

```
observeEvent( input$testingtrigger , {

  if (golem::app_dev()){
    file$original_file <- minifyr::minifyr_example(
      ext = input$testingtext
    )
    file$guess_minifier()
    file$type <- input$upload$type
    file$minified_file <- NULL
    file$original_name <- input$upload$name
    gargoyle::trigger("uploaded")
  }

})
```

We use this pattern so that we can combine it with a testing suite with {crrry}, using the following pattern:

```
test <- crrry::CrrryProc$new(
  chrome_bin = pagedown::find_chrome(),
  # Process to launch locally
  fun = "golem::document_and_reload();run_app()",
  # Note that you will need httpuv >= 1.5.2 for randomPort
  chrome_port = httpuv::randomPort(),
  headless = FALSE
)

test$wait_for_shiny_ready()
```

```
ext <- c("css", "js", "json", "html")
for (i in 1:length(ext)){
  # Set the extension value
  test$shiny_set_input("left_ui_1-testingtext", ext[i])
  # Trigger the file to be read
  test$shiny_set_input("left_ui_1-testingtrigger", i)
  # Launch the minification
  test$shiny_set_input("left_ui_1-launch", i)
}

test$stop()
```

It's safer to wrap these tests between if(interactive()), as running the checks outside of your current session might not launch the app correctly, and launching external processes (the one running the app with Chrome) might fail when run non-interactively. And on top of that, running these inside your CI might cause some pain, and of course, it will not work on CRAN checks.

We'll also be building "standard" function checks, which you can find in the test/ folder.

This step is available at github.com/ColinFay/minifying, folder step-4-strengthen[21].

Step 5: Deploy

As an example, we will deploy this app in three media: as a package, on RStudio Connect, and with Docker.

Before deployment checklist

☒ *devtools::check(), run from the command line, return 0 errors, 0 warnings, 0 notes*

☒ *The current version number is valid, i.e if the current app is an update, the version number has been bumped: it makes sense, before the first deployment, to keep a version number of 0.0.0.9000, and increment this dev version whenever we implement changes or do test deployments. Because this is a "true" deployment, we bumped the version to 0.1.0.*

---

[21]https://github.com/ColinFay/minifying/tree/master/step-4-strengthen

⊠ *Everything is fully documented*: we have documented all the functions, even the internal, there is a Vignette that describes the business logic, and the README is filled.

⊠ *Test coverage is good, i.e. you cover a sufficient amount of the codebase, and these tests cover the core/strategic algorithms.*

⊠ *The person to call if something goes wrong is clear to everyone involved in the product..*

⊠ *The debugging process is clear to everyone involved in the project, including how to communicate bugs to the developer team, and how long it will take to get changes implemented*: this project will be made open source, so the bug will have to be listed on the Github repo. To help that, we added a link to the GitHub repository on the application.

⊠ *(If relevant) The server it is deployed on has all the necessary software installed (Docker, Connect, Shiny Server, etc.) to make the application run.*

⊠ *The server has all the system requirements needed (the system libraries), and if not, they are installed with the application (if it's dockerized)*: NodeJS will need to be installed on Docker and on the server running RStudio Connect. A check is also added on top of `run_app()` for the availability of NodeJS on the system, especially for people installing it as a package. This check will also check if `node-minify` has been installed, and if not, it will be installed. This check might take some time to run, but it will only be performed the first time the app is launched.

⊠ *The application, if deployed on a server, will be deployed on a port which will be accessible by the users*: when building the Dockerfile using {golem}, the correct port is exposed (*i.e* the app will run on port 80, which is also made available). For the other medium, the port will be automatically chosen, either by {shiny} or by Connect.

⊠ *(If relevant) The environment variables from the production server are managed inside the application*: not relevant.

⊠ *(If relevant) The app is launched on the correct port, or at least this port can be configured via an environment variable*: not relevant.

⊠ *(If relevant) The server where the app is deployed has access to the data sources (database, API...)*: not relevant.

⊠ *If the app records data and there are backups for these data*: not relevant

Deploy as a tar.gz

To share an application as a tar.gz, you can call `devtools::build()`, which will compile a `tar.gz` file inside the parent folder of your current application. You can then share this archive, and install it with `remotes::install_local("path/to/tar.gz")`. Note that this can also be done with base R, but {remotes} offers a smarter way when it comes to managing the dependencies of your archived package.

This `tar.gz` can also be sent to a package repository; be it the CRAN or any other package manager you might have in your company.

Deploy on RStudio Connect

Once we are sure that the server running Connect has NodeJS installed, and that we have installed the minify module with `minifyr::minifyr_npm_install()`, we can create the app.R using `golem::add_rstudioconnect_file()`, and then push to the Connect server.

Deploy with Docker

To create the Dockerfile, we'll start by launching `golem::add_dockerfile()`. This function will compute the system requirements,[22] and create a generic `Dockerfile` for your application. Once this is done, we will create/update the `.dockerignore` file at the root of the package, so that unwanted files are not bundled with our Docker image.

Inside our `Dockerfile`, we will also change the default repo to use `https://packagemanager.rstudio.com/all/latest`, which proposes precompiled packages for our system, making the installation faster. We will also add an installation of NodeJS, which is needed by our application..

Then, we can go to our terminal and compile the image!

```
docker build -t minifying .
```

Now we've got a working image! We can try it with:

---

[22]Note that at the time of writing these lines, there is also an issue with the dependencies collected by the `sysreq` API, leading to an issue when attempting to compile the Dockerfile. Removing the installation of `libgit2-dev` solved the issue.

```
docker run -p 2811:80 minifying
```

This step is available at github.com/ColinFay/minifying, folder step-5-deploy[23].

---

[23]https://github.com/ColinFay/minifying/tree/master/step-5-deploy

Appendix B - Session Info

The current version of this book has been compiled on:

```
Sys.Date()
```

```
[1] "2021-07-16"
```

with the following configuration:

```
xfun::session_info()
```

```
R version 4.1.0 (2021-05-18)
Platform: x86_64-apple-darwin17.0 (64-bit)
Running under: macOS Mojave 10.14.6, RStudio 1.4.1717

Locale: en_US.UTF-8 / en_US.UTF-8 / en_US.UTF-8
/ C / en_US.UTF-8 / en_US.UTF-8

Package version:
  AsioHeaders_1.16.1.1
  askpass_1.1
  assertthat_0.2.1
  attachment_0.2.1
  attempt_0.3.1
  backports_1.2.1
  bank_0.0.0.9000
  base64enc_0.1.3
  beeswarm_0.4.0
  bench_1.1.1
  BH_1.75.0.0
  BiocManager_1.30.16
  bit_4.0.4
  bit64_4.0.5
```

```
blob_1.2.1
bookdown_0.22
brew_1.0.6
brio_1.1.2
bslib_0.2.5.1
cachem_1.0.5
callr_3.7.0
charlatan_0.4.0
cli_3.0.0
clipr_0.7.1
clisymbols_1.2.0
cloc_0.3.5
codetools_0.2-18
colorspace_2.0-2
colourvalues_0.3.7
commonmark_1.7
compiler_4.1.0
config_0.3.1
covr_3.5.1
cpp11_0.3.1
cranlogs_2.1.1
crayon_1.4.1
credentials_1.3.0
crosstalk_1.1.1
curl_4.3.2
cyclocomp_1.1.0
data.table_1.14.0
datasets_4.1.0
DBI_1.1.1
debugme_1.1.0
desc_1.3.0
devtools_2.4.2
dichromat_2.0-0
diffobj_0.3.4
digest_0.6.27
dockerfiler_0.1.3
dockerstats_0.1.0
dplyr_1.0.7
DT_0.18
dygraphs_1.1.1.6
ellipsis_0.3.2
evaluate_0.14
fakir_0.2.0.9000
fansi_0.5.0
farver_2.1.0
```

```
fastmap_1.1.0
fs_1.5.0
future_1.21.0
gargoyle_0.0.1
generics_0.1.0
gert_1.3.1
ggbeeswarm_0.6.0
ggplot2_3.3.5
gh_1.3.0
git2r_0.28.0
gitcreds_0.1.1
globals_0.14.0
glue_1.4.2
golem_0.3.1
golex_0.0.0.9000
goodpractice_1.0.2.9000
graphics_4.1.0
grDevices_4.1.0
grid_4.1.0
gridExtra_2.3
gtable_0.3.0
here_1.0.1
highr_0.9
hms_1.1.0
htmltools_0.5.1.1
htmlwidgets_1.5.3
httpuv_1.6.1
httr_1.4.2
ini_0.3.1
isoband_0.2.5
jquerylib_0.1.4
jsonlite_1.7.2
knitr_1.33
labeling_0.4.2
later_1.2.0
lattice_0.20-44
lazyeval_0.2.2
lifecycle_1.0.0
lintr_2.0.1
listenv_0.8.0
liteq_1.1.0
lubridate_1.7.10
magrittr_2.0.1
markdown_1.1
MASS_7.3.54
```

```
matlab_1.0.2
Matrix_1.3.3
memoise_2.0.0
methods_4.1.0
mgcv_1.8.35
mime_0.11
minifyr_0.0.0.9100
munsell_0.5.0
namer_0.1.5
nlme_3.1.152
openssl_1.4.4
packageMetrics2_1.0.1.9000
packrat_0.6.0
parallel_4.1.0
parallelly_1.26.1
parsedate_1.2.1
pillar_1.6.1
pkgbuild_1.2.0
pkgconfig_2.0.3
pkgload_1.2.1
plogr_0.2.0
plotly_4.9.4.1
praise_1.0.0
prettyunits_1.1.1
processx_3.5.2
profmem_0.6.0
progress_1.2.2
promises_1.2.0.1
ps_1.6.0
purrr_0.3.4
R.cache_0.15.0
R.methodsS3_1.8.1
R.oo_1.24.0
R.utils_2.10.1
R6_2.5.0
rappdirs_0.3.3
rcmdcheck_1.3.3
RColorBrewer_1.1.2
Rcpp_1.0.7
readr_1.4.0
rematch2_2.1.2
remotes_2.4.0
renv_0.12.2
rex_1.2.0
rlang_0.4.11
```

```
rmarkdown_2.9
roxygen2_7.1.1
rprojroot_2.0.2
rsconnect_0.8.18
RSQLite_2.2.7
rstudioapi_0.13
rversions_2.1.1
rvest_1.0.0
sass_0.4.0
scales_1.1.1
selectr_0.4.2
sessioninfo_1.1.1
shinipsum_0.0.0.9000
shiny_1.6.0
shinyloadtest_1.1.0
sourcetools_0.1.7
splines_4.1.0
stats_4.1.0
stringi_1.6.2
stringr_1.4.0
styler_1.5.1
svglite_2.0.0
sys_3.4
systemfonts_1.0.2
testthat_3.0.4
tibble_3.1.2
tictoc_1.0.1
tidyr_1.1.3
tidyselect_1.1.1
tidytuesday201942_0.0.0.9000
tinytex_0.32
tools_4.1.0
tzdb_0.1.1
usethis_2.0.1
utf8_1.2.1
utils_4.1.0
vctrs_0.3.8
vipor_0.4.5
viridis_0.6.1
viridisLite_0.4.0
vroom_1.5.2
waldo_0.2.5
websocket_1.4.0
whisker_0.4
whoami_1.3.0
```

```
withr_2.4.2
xfun_0.24
xml2_1.3.2
xmlparsedata_1.0.5
xopen_1.0.0
xtable_1.8-4
xts_0.12.1
yaml_2.2.1
yesno_0.1.2
zip_2.2.0
zoo_1.8-9
```

Bibliography

Allaire, J. (2020). *config: Manage Environment Specific Configuration Values.* R package version 0.3.1.

Attali, D. (2020). *shinyjs: Easily Improve the User Experience of Your Shiny Apps in Seconds.* R package version 2.0.0.

Attali, D. and Edwards, T. (2020). *shinyalert: Easily Create Pretty Popup Messages (Modals) in Shiny.* R package version 2.0.0.

Bengtsson, H. (2020a). *future: Unified Parallel and Distributed Processing in R for Everyone.* R package version 1.21.0.

Bengtsson, H. (2020b). *profmem: Simple Memory Profiling for R.* R package version 0.6.0.

Bengtsson, H. (2021). *R.cache: Fast and Light-Weight Caching (Memoization) of Objects and Results to Speed Up Computations.* R package version 0.15.0.

Billestrup, J., Stage, J., Bruun, A., Nielsen, L., and Nielsen, K. S. (2014). Creating and using personas in software development: Experiences from practice. In *Human-Centered Software Engineering*, pages 251–258. Springer Berlin Heidelberg.

Boettiger, C. and Eddelbuettel, D. (2017). An Introduction to Rocker: Docker Containers for R. *The R Journal*, 9(2):527–536.

Chang, W. (2021). *chromote: Headless Chrome Web Browser Interface.* R package version 0.0.0.9003.

Chang, W., Cheng, J., Allaire, J., Sievert, C., Schloerke, B., Xie, Y., Allen, J., McPherson, J., Dipert, A., and Borges, B. (2021a). *shiny: Web Application Framework for R.* R package version 1.6.0.

Chang, W., Csárdi, G., and Wickham, H. (2021b). *shinytest: Test Shiny Apps.* R package version 1.5.0.

Chang, W., Luraschi, J., and Mastny, T. (2020). *profvis: Interactive Visualizations for Profiling R Code.* R package version 0.3.7.

Chapman, C. N., Love, E., Milham, R. P., ElRif, P., and Alford, J. L. (2008). Quantitative evaluation of personas as information. *Proceedings of the Human Factors and Ergonomics Society Annual Meeting*, 52(16):1107–1111.

Cheng, J. (2021). *promises: Abstractions for Promise-Based Asynchronous Programming*. R package version 1.2.0.1.

Christian, B. and Griffiths, T. (2016). *Algorithms to Live by : The Computer Science of Human Decisions*. Henry Holt.

Coene, J. (2020). *fullPage: Three Families of Functions for Three Different Single Page Shiny Apps*. R package version 0.1.0.

Coene, J. (2021). *sever: Customise Shiny Disconnected Screens and Error Messages*. R package version 0.0.7.

Collado-Torres, L., Maynard, K. R., and Jaffe, A. E. (2020). *LIBD Visium spatial transcriptomics human pilot data inspector*. https://github.com/LieberInstitute/spatialLIBD - R package version 0.99.15.

Cooley, D. (2020). *geojsonsf: GeoJSON to Simple Feature Converter*. R package version 2.0.1.

Csardi, G. (2016). *cyclocomp: Cyclomatic Complexity of R Code*. R package version 1.1.0.

Csardi, G. (2021). *packageMetrics2: Collect Metrics about R Packages*. R package version 1.0.1.9000.

Csárdi, G. (2019a). *liteq: Lightweight Portable Message Queue Using SQLite*. R package version 1.1.0.

Csárdi, G. (2019b). *rcmdcheck: Run R CMD check from R and Capture Results*. R package version 1.3.3.

Csárdi, G. and Chang, W. (2021). *processx: Execute and Control System Processes*. R package version 3.5.2.

Csárdi, G., Müller, K., and Hester, J. (2021). *desc: Manipulate DESCRIPTION Files*. R package version 1.3.0.

Dowle, M. and Srinivasan, A. (2021). *data.table: Extension of 'data.frame'*. R package version 1.14.0.

Eddelbuettel, D., Francois, R., Allaire, J., Ushey, K., Kou, Q., Russell, N., Ucar, I., Bates, D., and Chambers, J. (2021). *Rcpp: Seamless R and C++ Integration*. R package version 1.0.7.

Eddelbuettel, D. and Knapp, B. (2021). *RcppSimdJson: Rcpp Bindings for the simdjson Header-Only Library for JSON Parsing*. R package version 0.1.5.

Fay, C. (2019). *dockerfiler: Easy Dockerfile Creation from R*. R package version 0.1.3.

Fay, C. (2020). *attempt: Tools for Defensive Programming*. R package version 0.3.1.

Fay, C. (2021a). *bubble: Launch and Interact with a NodeJS Session*. R package version 0.0.0.9003.

Fay, C. (2021b). *crrry: crrri recipes for shiny*. R package version 0.0.0.9001.

Fay, C. (2021c). *dockerstats: A Wrapper Around docker stats*. R package version 0.1.0.

Fay, C. (2021d). *gargoyle: An Event-Based Mechanism for Shiny*. R package version 0.0.1.

Fay, C. (2021e). *glouton: JS-cookies for Shiny*. R package version 0.0.0.9000.

Fay, C. (2021f). *hexmake: Hex Stickers Maker*. R package version 0.1.3.

Fay, C. (2021g). *minifyr: Minify CSS, HTML and JavaScript Files*. R package version 0.0.0.9100.

Fay, C. (2021h). *nessy: A NES css for Shiny*. R package version 0.0.0.9001.

Fay, C. (2021i). *resume: Bootstrap Resume Template for Shiny*. R package version 0.0.0.9000.

Fay, C. (2021j). *skeleton: Skeleton CSS for Shiny*. R package version 0.0.0.9000.

Fay, C. (2021k). *tidytuesday201942: A golem App for Tidy Tuesday*. R package version 0.0.0.9000.

Fay, C., Guyader, V., Rochette, S., and Girard, C. (2021). *golem: A Framework for Robust Shiny Applications*. R package version 0.3.1.

Fay, C. and Rochette, S. (2021a). *fakir: Create Fake Data in R for tutorials*. R package version 0.2.0.9000.

Fay, C. and Rochette, S. (2021b). *shinipsum: Lorem-Ipsum-like Helpers for fast Shiny Prototyping*. R package version 0.0.0.9000.

Garnier, S. (2021). *viridis: Colorblind-Friendly Color Maps for R*. R package version 0.6.1.

Gillespie, C. and Lovelace, R. (2017). *Efficient R programming*. O'Reilly Media, Inc, USA.

Granjon, D. (2021a). *bs4Dash: A Bootstrap 4 Version of shinydashboard*. R package version 2.0.0.

Granjon, D. (2021b). *shinydashboardPlus: Add More AdminLTE2 Components to shinydashboard*. R package version 2.0.1.

Granjon, D., Perrier, V., and Rudolf, I. (2021). *shinyMobile: Mobile Ready shiny Apps with Standalone Capabilities.* R package version 0.8.0.

Henry, L. and Wickham, H. (2020). *purrr: Functional Programming Tools.* R package version 0.3.4.

Hester, J. (2020a). *bench: High Precision Timing of R Expressions.* R package version 1.1.1.

Hester, J. (2020b). *covr: Test Coverage for Packages.* R package version 3.5.1.

Hester, J., Csárdi, G., Wickham, H., Chang, W., Morgan, M., and Tenenbaum, D. (2021). *remotes: R Package Installation from Remote Repositories, Including GitHub.* https://remotes.r-lib.org, https://github.com/r-lib/remotes.

Hester, J. and Wickham, H. (2021). *vroom: Read and Write Rectangular Text Data Quickly.* R package version 1.5.2.

Kim, G. (2016). *The DevOps Handbook: How to Create World-Class Agility, Reliability, and Security in Technology Organizations.* IT Revolution Press.

Kim, G. (2019). *The unicorn project: a novel about developers, digital disruption, and thriving in the age of data.* IT Revolution Press.

Krug, S. (2014). *Don't Make Me Think, Revisited: A Common Sense Approach to Web Usability (3rd Edition) (Voices That Matter).* New Riders.

Larbaoui, M. (2021). *tidymodules: A robust framework for developing shiny modules.* R package version 0.1.1.

Lemaire, M. (2020). *Refactoring At Scale.* Henry Holt.

Lesur, R. and Dervieux, C. (2021). *crrri: An Interface with Headless Chromium/Chrome.* https://rlesur.github.io/crrri/, https://github.com/RLesur/crrri.

Lumley, T. (2013). *dichromat: Color Schemes for Dichromats.* R package version 2.0-0.

Merlino, A. and Howard, P. (2020). *shinyFeedback: Display User Feedback in Shiny Apps.* R package version 0.3.0.

Müller, K. (2020). *here: A Simpler Way to Find Your Files.* R package version 1.0.1.

Müller, K., Wickham, H., James, D. A., and Falcon, S. (2021). *RSQLite: SQLite Interface for R.* R package version 2.2.7.

Nüst, D., Eddelbuettel, D., Bennett, D., Cannoodt, R., Clark, D., Daroczi, G., Edmondson, M., Fay, C., Hughes, E., Kjeldgaard, L., Lopp, S., Marwick, B., Nolis, H., Nolis, J., Ooi, H., Ram, K., Ross, N., Shepherd, L., Sólymos, P., Swetnam, T. L., Turaga, N., Petegem, C. V., Williams, J., Willis, C., and Xiao, N. (2020). The rockerverse: Packages and applications for containerization with r.

Ooms, J. (2014). The jsonlite package: A practical and consistent mapping between json data and r objects. *arXiv:1403.2805 [stat.CO]*.

Pebesma, E. (2021). *sf: Simple Features for R*. R package version 1.0-0.

R Core Team (2020). *foreign: Read Data Stored by Minitab, S, SAS, SPSS, Stata, Systat, Weka, dBase, ...* R package version 0.8-81.

R Special Interest Group on Databases (R-SIG-DB), Wickham, H., and Müller, K. (2021). *DBI: R Database Interface*. R package version 1.1.1.

Raymond, E. S. (2003). *The Art of UNIX Programming (The Addison-Wesley Professional Computng Series)*. Addison-Wesley.

Rochette, S. and Guyader, V. (2021). *attachment: Deal with Dependencies*. R package version 0.2.1.

Roebuck, P. (2014). *matlab: MATLAB emulation package*. R package version 1.0.2.

Rudis, B. and Danial, A. (2020). *cloc: Count Lines of Code, Comments and Whitespace in Source Files and Archives*. R package version 0.3.5.

Sakamoto, Y., Matsumoto, S., Tokunaga, S., Saiki, S., and Nakamura, M. (2015). Empirical study on effects of script minification and HTTP compression for traffic reduction. In *2015 Third International Conference on Digital Information, Networking, and Wireless Communications (DINWC)*. IEEE.

Schloerke, B., Dipert, A., and Borges, B. (2021). *shinyloadtest: Load Test Shiny Applications*. R package version 1.1.0.

Sidi, J. (2021). *covrpage: Create a Readme Summary Page for GitHub testthat and covr Outputs*. R package version 0.1.

Sidi, J. and Müller, K. (2019). *whereami: Reliably Return the Source and Call Location of a Command*. R package version 0.1.9.

Sievert, C. (2020). *Interactive Web-Based Data Visualization with R, plotly, and shiny*. Chapman and Hall/CRC.

Ushey, K. (2020). *renv: Project Environments*. R package version 0.12.2.

Vaidyanathan, R., Xie, Y., Allaire, J., Cheng, J., Sievert, C., and Russell, K. (2020). *htmlwidgets: HTML Widgets for R*. R package version 1.5.3.

Wickham, H. (2019). *Advanced R, Second Edition*. Chapman and Hall/CRC.

Wickham, H. (2021). *testthat: Unit Testing for R*. R package version 3.0.4.

Wickham, H., Averick, M., Bryan, J., Chang, W., McGowan, L. D., François, R., Grolemund, G., Hayes, A., Henry, L., Hester, J., Kuhn, M., Pedersen, T. L., Miller, E., Bache, S. M., Müller, K., Ooms, J., Robinson, D., Seidel, D. P., Spinu, V., Takahashi, K., Vaughan, D., Wilke, C., Woo, K., and Yutani, H. (2019). Welcome to the tidyverse. *Journal of Open Source Software*, 4(43):1686.

Wickham, H. and Bryan, J. (2019). *readxl: Read Excel Files*. R package version 1.3.1.

Wickham, H. and Bryan, J. (2020). *R Packages*.

Wickham, H. and Bryan, J. (2021). *usethis: Automate Package and Project Setup*. R package version 2.0.1.

Wickham, H., Chang, W., Henry, L., Pedersen, T. L., Takahashi, K., Wilke, C., Woo, K., Yutani, H., and Dunnington, D. (2021a). *ggplot2: Create Elegant Data Visualisations Using the Grammar of Graphics*. R package version 3.3.5.

Wickham, H., Danenberg, P., Csárdi, G., and Eugster, M. (2020). *roxygen2: In-Line Documentation for R*. R package version 7.1.1.

Wickham, H., François, R., Henry, L., and Müller, K. (2021b). *dplyr: A Grammar of Data Manipulation*. R package version 1.0.7.

Wickham, H., Girlich, M., and Ruiz, E. (2021c). *dbplyr: A dplyr Back End for Databases*. R package version 2.1.1.

Wickham, H. and Grolemund, G. (2017). *R for Data Science: Import, Tidy, Transform, Visualize, and Model Data*. O'Reilly Media.

Wickham, H. and Hester, J. (2020). *pkgbuild: Find Tools Needed to Build R Packages*. R package version 1.2.0.

Wickham, H., Hester, J., and Chang, W. (2021d). *devtools: Tools to Make Developing R Packages Easier*. R package version 2.4.2.

Wickham, H., Hester, J., Chang, W., Müller, K., and Cook, D. (2021e). *memoise: Memoisation of Functions*. R package version 2.0.0.

Wickham, H. and Miller, E. (2021). *haven: Import and Export SPSS, Stata and SAS Files*. R package version 2.4.1.

Woo, K., Kauer, N., and Montgomery, K. (2020). *dccvalidator: Metadata Validation for Data Coordinating Centers*. R package version 0.3.0.

Xie, Y. (2021a). *bookdown: Authoring Books and Technical Documents with R Markdown*. R package version 0.22.

Xie, Y. (2021b). *knitr: A General-Purpose Package for Dynamic Report Generation in R*. R package version 1.33.

Xie, Y., Cheng, J., and Tan, X. (2021). *DT: A Wrapper of the JavaScript Library DataTables*. R package version 0.18.

Index

Milton Keynes UK
Ingram Content Group UK Ltd.
UKHW031536071024
449327UK00024B/1878